CMOS Integrated Switching Power Converters

Gerard Villar Piqué · Eduard Alarcón

CMOS Integrated Switching Power Converters

A Structured Design Approach

Springer

Gerard Villar Piqué
NXP Semiconductors
Eindhoven
Netherlands
gerard.villar.pique@nxp.com

Eduard Alarcón
Technical University of Catalunya
Barcelona
Spain
ealarcon@eel.upc.edu

ISBN 978-1-4419-8842-3   e-ISBN 978-1-4419-8843-0
DOI 10.1007/978-1-4419-8843-0
Springer New York Dordrecht Heidelberg London

Library of Congress Control Number: 2011925259

Printed on acid-free paper

Springer is part of Springer Science+Business Media (www.springer.com)

Agraïments

Tot i ser l'autor d'aquest treball, és necessari que, en aquest punt, recordi i faci esment de tots aquells que d'una manera o altra m'han ajudat al llarg del camí que m'ha dut a obtenir el treball que teniu a les vostres mans.

En primer lloc, i per la seva implicació més directa i evident, cal que expressi la meva gratitud a l'**Eduard Alarcón** per una magnífica direcció i excel·lent guia científica, especialment durant els primers anys que he dedicat a elaborar aquesta tesi. D'entre les seves múltiples qualitats, m'agradaria destacar la seva visió d'alt nivell sobre cap a on cal encaminar les tasques de recerca, i també el seu tracte cordial, humà i proper.

També de forma molt especial, m'agradaria expressar el meu profund agraïment a en **Francesc Guinjoan** i a l'**Albert Poveda** pel seu recolzament tant científic com humà. Són moltes les coses bones que podria dir d'aquestes persones, però resumidament els considero els principals responsables de l'exemplar forma de treballar del grup de recerca en el que he tingut la sort de participar. Una forma de treballar que té en compte no només la dimensió tècnica de les persones, sinó també la humana. Una forma de treballar que recorda que el doctorat també ha de començar per una part de formació, i no pas per l'exigència ràpida dels resultats. Una forma de treballar que prima la qualitat de la recerca per sobre de la quantitat. I malauradament, una forma de treballar que, després de passar 6 anys vinculat al món de la recerca, m'ha semblat força menys habitual del que seria desitjable. Tot això és el que, en la meva opinió, és un dels principals actius i puntals del grup de recerca, i crec que cal no deixar-ho perdre, davant d'un entorn on es sol valorar sobretot la producció de resultats, oblidant les persones que hi estan implicades. Òbviament, per tot això, també m'agradaria estendre el meu agraïment a la resta de membres del grup de recerca.

També ha estat de capital importància l'ajut d'en **Jordi Madrenas**, sense la desinteressada col·laboració del qual res d'això no hauria estat possible.

Arribat a aquest punt, és el moment d'expressar el meu més sincer agraïment a en **Juan Negroni**, en **Felipe Osorio**, l'**Eduardo Aldrete**, la **Carolina Mora**, en **Lázaro Marco**, l'**Oriol Torres**, en **Guillermo Bedoya**, en **Jordi Ricart**, en **David Molinero**, en **Santi Pérez** i a la resta dels companys que han compartit amb mi aquesta empresa que implica fer el doctorat. Si des del punt de vista científico-tècnic, la fluïda i desencartonada interacció amb totes aquestes persones ha estat d'una inestimable ajuda; des d'un punt de vista personal, el guany en l'agradable ambient de treball ha estat molt superior. De fet, és amb aquestes persones amb qui comparteixes, al llarg d'innombrables cafès, les inquietuds, les decepcions i les alegries que inevitablement reculls al llarg de tot el doctorat. En realitat, crec que poques persones més, per la seva proximitat, arriben a fer-se càrrec de tot el que implica endinsar-se en aquesta tasca. I tot sovint, et recorden que també cal tenir cura dels altres aspectes de la vida, molts dels quals he pogut compartir amb ells (novament, en molts cafès i sopars). En definitiva, en el meu cas, he de dir que sempre n'he rebut recolzament, ànims i comprensió.

No cal dir que, en tot això, hi ha tingut un paper fonamental **la meva família**. El seu recolzament i comprensió constants i tot sovint la seva necessària paciència han resultat un suport imprescindible al llarg de tot aquest temps.

Finalment, i no per això menys important, vull expressar el meu agraïment a la resta dels meus **amics** que durant tot aquest temps tants bons moments han compartit amb mi.

Acknowledgements

In spite of being the author of this work, it is necessary that, at this point, I mention and remember all those people that, in any way, have helped me along the way that has lead me to obtain the work that you have in your hands.

In first place and for his more direct and evident implication, it is necessary that I express my gratitude to **Eduard Alarcón** for a magnificent direction and an excellent scientific guide, especially during the first years that I have devoted to elaborating this thesis. Among his multiple qualities, I would like to highlight his high level vision towards where it is necessary to direct the tasks of research, and also his cordial, human and next treatment.

Also in a very special way, I would like to express my gratitude to **Francesc Guinjoan** and **Albert Poveda** for their scientific and personal support. There are many good things that I could say about these persons, but summarizedly I consider them the main responsibles for the exemplary way of working of the research group in which I have been lucky to work with. A way of working that not only takes into account the technical dimension of the persons, but also the human one. A way of working which remembers that the doctorate should also start with an education cycle, and not for the fast demand of results. A way of working that incentives the quality of the research above its quantity. And unfortunately, a way of working that, after spending 6 years immersed in the world of scientific research, I have found quite rare of what it would be desirable. All these things compose what, in my opinion, is one of the main assets and backbones of this research group, and I believe that it is necessary not to let it lose, in front of an environment where the production of results is usually strengthened, forgetting the people involved. Obviously, because of all of this, I would also like to extend my gratitude to the rest of members of the research group.

The help of **Jordi Madrenas** has also been of capital importance, without the selfless collaboration of whom nothing of this would have been possible.

Arrived to this point, it is time to express my most sincere gratitude to **Juan Negroni, Felipe Osorio, Eduardo Aldrete, Carolina Mora, Lázaro Marco, Oriol Torres, Guillermo Bedoya, Jordi Ricart, David Molinero**, and **Santi Pérez** and to the rest of the colleagues that have shared with me this enterprise which implies to study the doctorate. Whether from the technical point of view, the fluent and disstiffened interaction with all these persons has been an inestimable support; from the personal point of view, the profit in the pleasant working environment has been very superior. As a matter of fact, they are the people who share with you, throughout innumerable coffee-breaks, the interests, the disappointments and the joys that you unavoidably encounter along the doctorate studies. In fact, I believe that few persons more, because of their proximity, really understand all what implies to get into this task. And often, they also remember you the need to take care of other faces of life, many of which I have been able to share with them (again, along many coffee-breaks and dinners). In short, in my case, I have to say that I have always received their support and understanding.

It is not necessary to say that, in all these things, **my family** has had a fundamental role. Their constant support and understanding and very often their necessary patience have become an indispensable support along this time.

Last but by no means least, I want to express my gratitude to the rest of **my friends** for so many good moments that they have shared with me.

Funding

- Partial funding by project TEC2004-05608-C02-01 from the Spanish MCYT and EU FEDER funds is acknowledged.
- Partial funding by project TIC-2001-2183 from the Spanish MCYT.

Contents

List of Figures

List of Tables

Table of Symbols

Symbol	Meaning
A_{Cc}	MOSCAP structure capacitive area
A_{C_o}	Output capacitor occupied area
A_{Cr}	MOSCAP structure routing area
A_L	Inductor occupied area
A_{MOS}	Power MOS occupied area
A_{driver}	Driver occupied area
A_{total}	Total occupied area
$\mathbf{B}$	Magnetic flux density
C	Integration capacitor in the dnN1 and dnN2 signals generation circuits (**Sect. 7.1**)
C_A	Total parasitic capacitance connected to the A-node
C_B	Total parasitic capacitance connected to the B-node
C_{MIM}	Capacitive density of Metal-Insulator-Metal capacitor
C_{PAD}	Bonding pad parasitic capacitor
C_T	Total gate capacitance of the tapered buffer
C_{V_x}	x-node total parasitic capacitor
C_{dd}	Total parasitic capacitor observed from the transistor drain
C_{gate}	MOS transistor gate capacitance
C_{in}	Tapered buffer input capacitance
C_{jd}	Drain junction parasitic capacitor
C_{mm}	Capacitive density of a metal-to-metal parasitic capacitor
C_n	Gate capacitance of a single MOSCAP cell
C_o	Converter output capacitor
C_{out}	Capacitance to be driven by the output of the tapered buffer
C_{ox}	MOS transistor oxid gate capacitance per area
$C_{poly-poly}$	Capacitive density of a polysilicon-polysilicon capacitor
C_{sd}	Transistor source-to-drain parasitic capacitance
C_{sg}	Transistor source-to-gate parasitic capacitance
C_x	3-level converter flying capacitor
D_1	d_1 switching signal duty cycle
D_3	d_3 switching signal duty cycle

E_D	Total energy switching losses related to the driver design. This includes the energy losses of the driver itself plus a fraction of the power MOS switching losses due to a non-instantaneous state transition.
E_{TRT_D}	Switching energy losses of a power MOS transistor due to a non-instantaneous transition of its gate voltage
E_{cond_T1}	Transistor conduction energy losses due to current flowing during T_1
E_{driver}	Energy consumed by the tapered buffer itself, in a complete switching cycle
E_{sw}	Switching energy losses
E_{sw_C}	Capacitive switching losses
E_{sw_R}	Resistive switching losses
ESR	Equivalent Series Resistance of any component, generally used for capacitors or inductors
ESR_T	MOSCAP structure total ESR
ESR_{T_min}	Minimum attainable value of the total MOSCAP structure ESR
ESR_{min}	Minimum attainable ESR of a single MOSCAP
$Error_{i_L}$	Error commited, in terms of the inductor current, at the moment of switching-off the NMOS power transistor
H_1	Variable that symbolizes the detection of a NMOS switching-off event, when the inductor current is greater than zero.
H_2	Variable that symbolizes the detection of a NMOS switching-off event, when the inductor current is lower than zero.
I_0	Inductor current initial value
I_{C_o}	Output capacitor current RMS value
I_{C_x}	C_x capacitor current RMS value
I_L	Inductor current RMS value
I_{LT_1RMS}	Inductor current RMS value considering that only conduction along the T_1 state exists
I_{LT_2RMS}	Inductor current RMS value considering that only conduction along the T_2 state exists
I_{LT_4RMS}	Inductor current RMS value considering that only conduction along the T_4 state exists
I_{L_max}	Inductor current maximum value
I_{L_min}	Inductor current minimum value
I_{Lsw}	Switching inductor current
I_{MOS}	MOS transistor current RMS value
I_{N_1}	N_1 transistor current RMS value
I_{N_2}	N_2 transistor current RMS value
I_{NMOS_CCM}	NMOS current RMS value in CCM operation
I_{NMOS_DCM}	NMOS current RMS value in DCM operation

I_{P_1}	P_1 transistor current RMS value
I_{P_2}	P_2 transistor current RMS value
I_{PMOS_CCM}	PMOS current RMS value in CCM operation
I_{PMOS_DCM}	PMOS current RMS value in DCM operation
I_{T_1}	Inductor current at the end of T_1
I_{T_4}	Inductor current at the end of T_4
I_o	Output current
I_{o_max}	Maximum output current
$I_{o\_max\_DCM}$	Maximum output current for which the power converter remains DCM operated
I_{off}	Transistor turn-off applied current
I_{on}	Transistor turn-on applied current
K_1	Fitting coefficient for square spiral inductor
K_2	Fitting coefficient for square spiral inductor
L	Self-inductance coefficient
L_{ij}	Mutual inductance between the ith and the jth coils
L_{ch}	Power transistor channel length
L_{min}	MOS transistor minimum channel length
L_{diff}	Transistor drain and source diffusion area length
N_1	NMOS power switch in the 3-level converter
N_2	NMOS power switch in the 3-level converter
P_1	PMOS power switch in the 3-level converter
P_2	PMOS power switch in the 3-level converter
P_{C_o}	Output capacitor resistive losses
P_{L_cond}	Inductor conduction losses
P_{cond}	Power-MOS conduction losses
P_{driver}	Driver power losses
P_{losses}	Converter total power losses
P_{out}	Converter output power
P_{sw}	Switching power losses
Q_A	Partial charge stored in the output capacitor of a 3-level converter
Q_B	Partial charge stored in the output capacitor of a 3-level converter
Q_1	Charge stored in the C_x capacitor during the T_1 state
Q_2	Charge stored in the C_x capacitor during the T_2 state
Q_4	Charge stored in the C_x capacitor during the T_4 state
Q_T	Total charge spent to change the state of the tapered buffer when its output is connected to the load device (including both state transitions).
Q_{e1}	Unitary-effort charge spent to change the state of a minimum sized inverter when is loaded with an identical inverter (including both state transitions)
Q_i	Intrinsic charge spent to change the state of a minimum sized inverter (including both state transitions)

Q_{out}	Charge spent from the voltage source to change the state (including both state transitions) of the load device connected to the tapered buffer output
R_1	Transistor channel resistance at the beginning of the turning-on action
R_2	Transistor channel resistance at the end of the turning-on action
R_{C_o}	Output capacitor equivalent series resistance (ESR)
R_G	Resistor corresponding to the gate terminal of a single MOS capacitor
R_L	Inductor equivalent series resistance (ESR)
R_{MET1}	*Metal1* layer square resistance
R_{MET2}	*Metal2* layer square resistance
R_{MET3}	*Metal3* layer square resistance
R_{MIM}	Metal-Insulator-Metal capacitor equivalent series resistance (ESR)
R_{TRT}	Transistor channel resistance
R_{bp}	Bonding pad resistance
R_{ch}	Resistor corresponding to the channel terminal of a single MOS capacitor
R_{cont}	Diffusion/Polysilicon-to-*metal1* contact resistance
R_{driver}	Driver equivalent output resistance
R_{ed}	Jin's model channel resistance due to changes in the channel charge
R_{on}	Power-MOS channel conduction resistance
R_{poly}	Polysilicon layer square resistance
R_s	Resistance value of the resistor used to sense the inductor current
R_{st}	Jin's model channel static channel resistance
R_{via1}	*Via1* resistance
R_{via2}	*Via2* resistance
$R_\square$	Inductor conductor square-resistance
T_1	3-level converter state characterized by the conduction of P_1 and N_2 power switches
T_1'	3-level converter inactivity state previous or subsequent to T_1
T_{1M}'	Time by which the inductor current crosses I_o in a 3-level converter, during T_1
T_{1m}'	Time by which the inductor current crosses I_o in a 3-level converter, during T_1
T_2	3-level converter state characterized by the conduction of N_1 and N_2 power switches
T_2'	Time by which the inductor current crosses I_o in a 3-level converter, during T_2
T_3	3-level converter state characterized by the conduction of P_2 and N_1 power switches
T_3'	3-level converter inactivity state previous or subsequent to T_2

T_4	3-level converter state characterized by the conduction of P_1 and P_2 power switches
T'_4	Time by which the inductor current crosses I_o in a 3-level converter, during T_4
T_i	Inactivity phase of a Buck converter in DCM operation.
T_{off}	NMOS conduction phase along a switching cycle of the Buck converter. It also means the duration of this phase.
T_{on}	PMOS conduction phase along a switching cycle of the Buck converter. It also means the duration of this phase.
T_s	Switching cycle duration
V_0	Initial voltage
$V_{C_o\_0}$	Output capacitor initial voltage (without considering its ESR)
$V_{TN/P}$	MOS transistor (N-type or P-type) threshold voltage
V_{bat}	Battery voltage (power converter input voltage)
V_d	Transistor drain voltage
V_{dd}	Voltage supply (usually is the same as the battery voltage)
V_{dd_driver}	Driver specific voltage supply
V_{ds}	Drain-to-source transistor voltage
V_{ds_off}	Drain-to-source transistor voltage when it is turned-off
V_f	Final voltage
V_{fb1}	Voltage used to adjust the T_2 state duration, to get Zero-Current-Switching operation, corresponding to the N1 switch
V_{fb2}	Voltage used to adjust the T_2 state duration, to get Zero-Current-Switching operation, corresponding to the N2 switch
V_{fb_BD1}	Voltage used to adjust the dead-time corresponding to the P1-N1 switches pair
V_{fb_BD2}	Voltage used to adjust the dead-time corresponding to the P2-N2 switches pair
V_{gs}	MOS transistor gate-source voltage
V_{max}	Maximum allowed gate-to-source/drain/body voltage of a MOSFET transistor
V_o	Output voltage
V_{o_max}	Output voltage maximum value
V_{o_min}	Output voltage minimum value
V_{off}	Transistor turn-off applied voltage
V_{offset}	Input offset voltage of a voltage comparator
V_{on}	Transistor turn-on applied voltage
V_{t_off}	Hysteretic control voltage threshold that determines the beginning of the T_{off} state
V_{t_on}	Hysteretic control voltage threshold that determines the beginning of the T_{on} state
V_{term}	Thermal voltage
V_{x_max}	High level of the v_x voltage

W_{ch}	MOS transistor channel width
W_{N_1}	N_1 transistor channel width
W_{N_2}	N_2 transistor channel width
W_{P_1}	P_1 transistor channel width
W_{P_2}	P_2 transistor channel width
W_{min}	MOS transistor minimum channel width
W_n	NMOS transistor channel width of a minimum inverter
W_{n_driver}	Channel width of the NMOS transistor of the last driver stage
W_p	PMOS transistor channel width of a minimum inverter
W_{power_MOS}	Power transistor channel width
$\frac{W}{L}\vert_{opt}$	Single MOSCAP optimum aspect ratio
a	Half of the triangle side length.
	Sect. 3.2, overhead area width of a single MOSCAP cell
a_D	Ratio used to define the tapering factor (e.g. C_{out}/C_{in})
b	Overhead area length of a single MOSCAP cell
b_D	Ratio that relates a fraction of the power MOSFET switching losses and the output fall-rise time of a tapered buffer
d_1	3-level converter switching signal
d_3	3-level converter switching signal
d_{N_1}	N_1 transistor gate switching signal
d_{N_2}	N_2 transistor gate switching signal
d_{P_1}	P_1 transistor gate switching signal
d_{P_2}	P_2 transistor gate switching signal
d_{avg}	Spiral inductor average diameter
d_{in}	Spiral inductor inner diameter
d_{out}	Spiral inductor outer diameter
f	Driver chain of inverters tapering factor
f_s	Switching frequency
f_{s_opt}	Optimum switching frequency
i_{C_o}	Output capacitor current instantaneous value
i_L	Inductor current instantaneous value
i_{MOS}	Transistor instantaneous current
k	Proportionality factor between the switching frequency and the output current
	Sect. 7.1, transconductance factor of a MOS transistor in the active region
l_L	Inductor conductor length
n	Number of inverters of the driver
	Sect. 3.2, MOSCAP structure number of cells
n_L	Spiral inductor number of turns
n_{opt}	Optimum number of inverters of a tapered buffer
n_R	Number of resistance values of the staggered approach used in the resistive switching losses evaluation
n_{sides}	Number of sides of a geometrical shape
p	Distance between two adjacent turns of a spiral inductor

q	Reduction of half of the side of the triangular spiral, for any inner coil
r	Inductor conductor radius
r_{opt}	MOSCAP structure optimum aspect ratio
s_{ext}	Triangle outer side length
t_{BD}	Dead-time between the turning-off and the turning-on of two complementary power switches of a power converter
t_d	Signal propagation delay of a chain of inverters
t_{d_comp}	Input-to-output delay of a voltage comparator
t_{di}	Intrinsic delay of a minimum sized inverter
t_{de1}	Unitary-effort delay of a minimum sized inverter when is loaded with an identical inverter
t_{fr}	Average fall-rise time of the tapered buffer output voltage
t_{fri}	Intrinsic fall-rise time of a minimum sized inverter
t_{fre1}	Unitary-effort fall-rise time of a minimum sized inverter when is loaded with an identical inverter
t_m	Inductor conductor thickness
$t_{sw\_on/off}$	Transistor turn-on (turn-off) duration time
v_A	A-node voltage
v_B	B-node voltage
v_{C_x}	C_x capacitor voltage
v_{gs}	Transistor gate-to-source instantaneous voltage
v_o	Output voltage instantaneous value
v_x	x-node instantaneous voltage
w_L	Inductor conductor width
w_{pn}	Ratio between PMOS and NMOS transistors channel widths of an inverter of a tapered buffer
Γ	Switching power converter global merit figure
Γ_C	MOSCAP structure merit figure
Γ_L	Inductor merit figure
Δv_o	Converter output voltage ripple
Δv_{C_x}	C_x capacitor voltage increment along T_1 and T_3 converter states
Φ	Magnetic flux
α	Jin's model gate resistance coefficient
	Sect. 7.1, exponent of the current expression of a MOS transistor in the active region, according to the α-power model
β	Jin's model external path resistance coefficient
γ	Jin's model channel resistance coefficient
γ_{LA}	Weight coefficient in inductor merit figure definition, related with occupied area
γ_{LR}	Weight coefficient in inductor merit figure definition, related with series resistance
δ	Skin-depth
δ_{cont}	Diffusion/Polysilicon-to-*metal1* contacts linear density
δ_{via1}	*Via1* linear density

δ_{via2}	*Via2* linear density
ζ	Bonding wire resistivity
η	Power converter energy efficiency
	Sect. 3.2, Jin's model channel resistance coefficient
η_{min}	Minimum value of the power converter energy efficiency
μ	Media permeability
μ_0	Permeability of vacuum
$\mu_{N/P}$	MOS transistor (N-type or P-type) majority carriers effective mobility
μ_r	Media relative permeability
ρ	Spiral inductor fill ratio
σ	Inductor conductor conductivity

Chapter 1
Introduction

1.1 Motivation

The application framework of the presented research work is stablished in the area of the low-power portable stand-alone devices. Considering as a paradigm application the implementation of future portable terminals for telecommunications, two opposed trends coexist that determine the need for an optimized energy management. An important increase in power consumption is expected in future battery operated terminals, specially for the third and fourth generation systems (3G, 4G), due to the increasing demand in their functionalities, extended towards audiovisual communication and a much higher multimedia data flow (the last issue could also be referred to the personal audio players). In addition to this, other low-power stand-alone devices, such as the telemetric sensing devices and wireless sensor networks motes, should be the application target of the herein presented work. Microsystems and nanoelectronic circuits power supply fit this category as well.

On the other hand, the energy density increase of power sources (nowadays based in lithium-ion batteries) is expected to evolve much slower than the mentioned power demands of future systems. Furthermore, an improvement in terms of ergonomics and compactness is ever demanded in this kind of devices, as well as a longer operating life (which strongly constraints the use and selection of the whole powering system, i.e. from the battery to the power management circuitry).

In front of this scenario, a highly optimized energy management is needed in order to preclude a powering crisis in this kind of systems. This optimized energy management requires investigation in two complementary subjects:

- Development of different techniques to reduce the energy consumption of load subsystems (digital circuits, processors and DSP, as well as RF circuits). A notable amount of work has been done in this area, resulting in some advanced techniques such as the Adaptive Voltage Scaling (AVS).
- Optimization of the power management subsystems (in terms of energy efficiency) between the power source and the on-chip load.

In the context of the later research line, it is observed that the continuous trend towards the miniaturization of energy control and management subsystems (strictly

G. Villar Piqué, E. Alarcón, *CMOS Integrated Switching Power Converters*,
DOI 10.1007/978-1-4419-8843-0_1, © Springer Science+Business Media, LLC 2011

required throughout any telecommunications and computation portable systems or in aerospace applications), which are in charge of providing the required supply voltage and guarantee an efficient energy conversion, lies on the global impact that those power management subsystems represent in terms of weight and volume occupation.

Therefore, the research on the powering processing systems have become a key issue in the roadmap of such applications, as it previously happened in the information processing systems area. As a response in front of all the aforementioned issues, the main specifications of the powering processing systems have become to provide high energy-efficiency, miniaturization, reproducibility, and the always present cost reduction (specially in mass market products). All these requirements determine a line of convergence towards the fully monolithic integration of the powering systems.

Moreover, other benefits would stem up from the fully integration of the power management systems: complete customization for any different design, and reduction of the effect of the several parasitic elements from the system package on the power supply conditioning.

In case of full integration, the compactness and the cost reduction appear as a joint target, since the occupied area of silicon is directly related to system implementation cost.

Having determined the interest of achieving the full monolithic integration of the power management systems, it is also of primordial relevance the technology used to develope them, in terms of the fabrication costs, performance, design complexity, availability, etc... In this sense, a range of technological options appears, as regards the semiconductor used as substrate or the devices developed: bipolar, standard CMOS (fully digital or mixed signal), complementary bipolar-CMOS (BiCMOS), Silicon-Germanium (SiGe), Gallium-Arsenide (GaAs), Silicon-On-Insulator (SOI).

Although many of them offer much better device performance, nowadays, the most widely used is the standard CMOS technology. In fact, due to its extended production (mainly in the field of digital electronics), the derived lower fabrication costs have driven the RF systems design from more suitable technologies (i.e. GaAs) to the standard CMOS, at the expenses of poorer performance, which habitually have pushed the designer skills to develope more ingenious solutions, in order to get the desired functionality. The same reasoning could be applied to the field of high-performance pure analog systems.

Consequently, being the standard CMOS technology the convergence line of the different design fields that appear in the Systems-On-Chip (SOC), it is proposed to use it to develope the integrated power management systems.

At this point, it is also necessary a brief introduction on several example applications that would benefit from the aforementioned on-chip power management.

- An energy efficient simple voltage regulation of the power source for typical circuits that can be found in a System-On-Chip.
- Variable voltage supply of RF power amplifiers. Particularly interesting is the technique of polar power amplification for RF amplifiers, known as *envelope*

elimination and restoration or Kahn technique, which requires a baseband and wideband amplifier, ideally implemented by a switching amplifier [8–11].

- The *Adaptive Voltage Scaling (AVS)* technique that leads to an optimization of the power consumption in digital circuits (microprocessors and DSP), by varying the voltage supply to the minimum required as a function of the computational load. Such approach, in an SOC environment, requires an integrated efficient on-chip converter, hence becoming one of the bottlenecks in the success of this power-saving technique [12–17].

Since the last two applications would require the design and development of specialized control systems, and their corresponding implementation, this work focuses on the first one, as a first step in the development of more complex on-chip power management systems.

Three different approaches can be found for active power regulation:

Linear converters, whose efficiency is in first-order proportional to the input-to-output voltage ratio, although they can be easily integrated because their lack of reactive components. Since they offer a ripple-free output voltage they are preferred as power sources for analog and RF systems. As observed, their main drawbacks are the poor power efficiency, and their incapability to provide output voltages higher than the input [18, 19].

Switched capacitors converters (charge-pumps). This kind of power converters can provide an output voltage higher than its input without the need of inductors (which is interesting in applications such as the supply of RF amplifiers and TFT displays). Since they only require capacitors and transistors, they can be easily integrated as well, but at the expense of using a larger amount of occupied area, as compared to the linear counterparts. Their energy efficiency is lower than in the LC converters, and the output regulation is rather poor [20–24].

LC switching power converters, that exhibit the highest efficiency (theoretically 100%), and depending on the topology, they provide step-down and step-up functionalities. Their main drawbacks are the existence of ripple in their output voltage (due to their switched operation), and the unavoidable requirement for at least one inductor, which becomes the most difficult component to integrate in standard CMOS. In the rest of this thesis, they will be simply referred as switching power converters [25–28].

The two first options are currently commercially available because of their better suitability to integration in spite of its lower efficiency. The third one, which is the main target of this thesis, constitutes the goal implementation in future systems because of its expected higher energy efficiency.

In spite of all the developed work related to the switching powers converters area, paradigm of the high efficiency power conversion, additional research is required to achieve their monolithic integration [29–36]. A complete success in this field, that is the conceptual framework in which the present work focuses, has not been accomplished although it is becoming especially strategic and many applications would benefit from it. This is due to the fact that it requires a multidisciplinary approach from several areas such as power electronics (particularly high-frequency switching power converters), control theory (linear and non-linear techniques) and analog microelectronics (in order to implement the corresponding control loops, as well as the required power components of the power converters). This necessary concurrence should eventually lead to the Powered-System-On-Chip (PSOC), which not only includes the particular information processing systems (this is, digital, analog and RF parts), but also includes the power management systems, so that the whole system is included on the same silicon die or, at least, in the same package.

1.2 Technological Scenario

After the presentation of the conceptual framework that surrounds the presented thesis (including the global interest, justification and requirements), in this section more particular information is provided to justify the selected application parameters used as the design guideline to propose the fully integration of the switching power converter.

In order to choose the main converter functionality (this is, input-to-output voltage ratio), it is important to have a look at some of the main data related to the most recent battery technologies (which, throughout the whole thesis, are supposed to be the input power source), and the CMOS technologies (which will be the target loads to be supplied by the designed switching power converter).

In Fig. 1.1 the energy density and the output voltage of the most widely used battery technologies (particularly, for mass market products), can be observed. It is noted that the batteries that offer the highest energy density are the *Lithium-ion* and the *Lithium-polymer*, that present a nominal output voltage of 3.6 and 3.7 V, respectively. Additionally, it is observed that the more recent *Zinc-Air* also offer a high energy density. However, they usage is much more restricted than the Lithium based batteries. The year of their introduction into the market has been noted between brackets.

As the output loads regards, it is interesting to observe the evolution of the voltage levels required to supply the cores of the most recent technology processes. In Fig. 1.2 this evolution is presented, while showing the year of the first commercial availability of each technological node. Although most of the data have been obtained from the *UMC foundry* webpage [3], the forecast for the 2010 and 2013 years is obtained from the *International Technology Roadmap for Semiconductors* webpage [4]. From the exposed data, the trend to reduce the voltage supply (targeting the consequent static and dynamic power consumption reduction) becomes clear, although it seems to saturate in the future processes (due to the non-zero V_T threshold of the MOS transistors).

From the previous technological considerations, it results that a wider use of the step-down power conversion is expected in the nearest future. Therefore, in

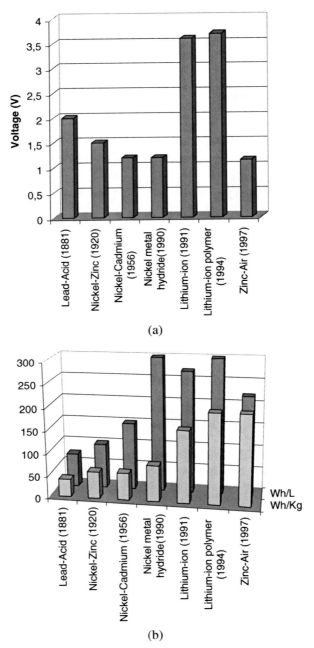

Fig. 1.1 Main battery characteristics: **a** cell voltage; **b** energy density (in terms of energy per weight or per volume). Data obtained from [1] and [2]

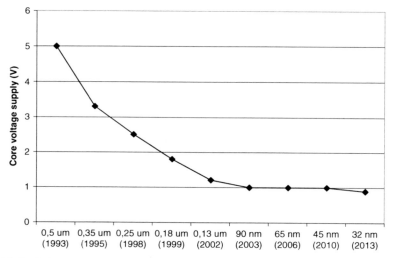

Fig. 1.2 Evolution of the voltage level required to supply the most recent and future standard CMOS technologies (referred to logic cores). Data was obtained from [3] and [4]

this thesis *it is proposed to develop a fully integrated step-down switching power converter*, whose nominal input and output voltages are $V_{bat} = 3.6$ V and $V_o = 1$ V, respectively.

Regarding the maximum allowable output ripple, it is set to $\Delta V_o = 50$ mV, following the 5% criterion.

Finally, the nominal output load current of the designed converter has been set to $I_o = 100$ mA, as a representative of the requirements of the low-power parts of portable devices.

In the development of the microelectronic designs presented along the whole thesis, the selected technological process has been determined to be the UMC 0.25 μm mixed-signal CMOS process.

1.3 Thesis Outline

After the introduction (Chap. 1), in Chap. 2 the *design space exploration* is introduced and proposed to be applied to the switching power converters design and optimization. Indeed, a global procedure including several optimizations for each component design, which afterwards are included in the global design space exploration, is proposed. In order to exemplify the presented design methodology, the same chapter covers the complete design of a Buck converter, where simplified models for each of the required components are used. The chapter also presents a complete analytical model for the output voltage ripple evaluation, when taking into account the output capacitor ESR.

Chapter 3 presents the detailed models and optimization procedures of the four different components that compose the Buck converter: inductor, capacitor, power drivers (tapered buffers) and power transistors. In all cases, a state of the art review is firstly exposed, suggesting the requirements uncovered by the previous works, which justifies the proposed implementations and models. While only in some cases (capacitor and inductor) a particular microelectronic implementation is proposed, for any of them, models focusing on the energy losses evaluation, as well as the occupied area, are provided.

From the detailed models (focused on a particular on-chip implementation) presented in the previous chapter, in Chap. 4 the complete design space exploration corresponding to the Buck converter is carried out. The obtained results from such optimization procedure, that yield a DCM operated converter, are discussed and found to offer too low performance. Consequently, the search for a different converter topology more suitable to its fully monolithic integration is exposed.

As a result, in Chap. 5, the implementation of an integrated 3-level converter (with a floating capacitor) as a voltage regulator is suggested and justified. As a first step, an analytical study of this topology is required in order to accurately determine the impact of its main design variables upon the most representative parameters and magnitudes, since the fully integration may force to extreme their values. Then, a particular bootstrap scheme to connect and supply the four required power drivers is proposed in order to allow the use of core transistors as power switches, which leads to improved power losses reduction. The suggested use of the 3-level topology requires the modification of some of the losses models presented in Chap. 3, which are explained in the last section of this chapter.

At this point, the design space exploration is carried out towards the design and optimization of the 3-level converter, whose results are collected in Chap. 6. First of all, the results from the optimization are analyzed and an important improvement on the classical Buck converter design is obtained (in this case, DCM operation is also found to be the most convenient for the focused application). However, the need to consider a wide range of output current values instead of just a single value is exposed and justified. Additionally, this implies a proportional switching frequency modulation, which in turn, produces an output ripple increase as the output current is reduced. The main consequence of this new consideration is a stronger constraint of the design space, which forces to carry out a new design space exploration, from which the 3-level converter design to be finally implemented is obtained.

Chapter 7 covers the microelectronic design and implementation of the circuits required to approach the converter switched operation as expected (mainly due to the employment of the synchronous rectification scheme in addition to the DCM operation). Therefore, in this chapter, full custom mixed-signal circuit proposals to automatically adjust the dead-time between power switches gates signals (in order to avoid the body-diode and the shot-through events), and the moment at which the NMOS switches are turn-off (in order to obtain Zero-Current-Switching operation), are presented and validated by means of transistor-level simulations. The design of other additional needed circuits is also covered in this chapter. In addition to this, the layout designs corresponding to all these circuits, as well as the converter power components (namely inductor, capacitors, power switches and power drivers) is also presented. At the end of the chapter, transistor-level simulation results involving the whole designed system are presented, and the proper functionality as well as good matching with the proposed models is observed.

Finally, Chap. 8 summarizes the conclusions of the thesis, also including the future research lines derived from the presented work.

Chapter 2
The Design Space Exploration: Simplified Case

Abstract In this chapter, the *design space exploration* is exposed and explained as the main tool to get an optimized design of a fully integrated switching power converter. The converter output voltage ripple is used to constraint the design space, while the merit figure to be optimized is defined as the power efficiency versus the occupied silicon area ratio. Additionally, an overall structure of the design space exploration including particular optimizations for each component design is proposed. Then, simplified models for the converter integrated components (i.e. inductor, capacitor, power drivers and power transistors) are presented, as well as an analytical output ripple model that takes into account the output capacitor *Equivalent-Series-Resistance* (ESR). Finally, the design space exploration optimization procedure is exemplified by means of the design of a classical Buck converter, where the simplified components models are used.

2.1 Main Concepts and Design Procedure

In the first steps of a switching power converter design (once the required topology has been determined), there are many questions and decisions to answer regarding input and output voltage ranges, output current, switching frequency, the reactive components values (mainly inductors and capacitors), transformer input–output voltage ratio (if any is required by the design), even some dynamic performance parameters such as bandwidth. If different technological options are considered to implement the power switches and their corresponding drivers, the design options can become overwhelming.

In order to develop an appropriate design procedure, it is interesting to classify the design variables or parameters into different categories to identify their impact upon the design characteristics or performance. In the following, groups proposed herein are presented.

Application parameters. Those parameters that are completely defined by the application needs, like input and output voltage ranges, output current capability and output voltage ripple. Usually, application parameters appear as *hard constraints* in the design space. Additionally, other less conventional

G. Villar Piqué, E. Alarcón, *CMOS Integrated Switching Power Converters*,
DOI 10.1007/978-1-4419-8843-0_2, © Springer Science+Business Media, LLC 2011

parameters or constrains can be defined: radiated interferences spectrum, maximum input current, etc...

Static design variables. Conventional design variables, such as inductance and capacitance value, or the voltage conversion ratio of a transformer. We call them 'static' because although they offer degrees of freedom in the design stage, they become fixed once the design is implemented.

Dynamic design variables. The main dynamic design variable that appears in the design of a conventional switching power converter is the switching frequency, since although it can be determined during the design stage, it offers an additional degree of freedom once the converter is implemented, that allows to better fit to application parameter changes.

Performance factor. This category includes those factors that indicate the power converter quality, once application parameters are satisfied or accomplished. Most common performance factors are energy efficiency in power conversion, the bandwidth of the output response versus the control signal, the volume and area occupation, its weight. They can appear as *soft constraints*, since they might not invalidate the design despite it could be interesting to maximize or minimize them. In fact, depending on the application, they could be totally or partially defined as *application parameters* (i.e. in case where a minimum power efficiency is required, or a maximum weight is allowed).

Technological information. This refers to information related to the physical implementation of the converter components, specially in case of the power switches and their corresponding drivers. Although first order approach designs do not require detailed information about the technology used to implement any component, information about non idealities and parasitics is needed to take into account some performance factors (e.g. energy efficiency).

In Table 2.1, the previous classification is summarized. In this case, only those variables that are particularly relevant for the case of integrated switching converters are shown.

Obviously, the categories and what they include could change if non conventional designs are considered, i.e. multiple conversion ratio transformers, variable capacitors, etc...

From the previous classification a design procedure is derived, which is described in the following.

1. First of all, taking into account the application parameters, some converter topologies are selected to fit specifications, and the corresponding required ideal analytical expressions for voltages and currents are obtained. In this design step, system-level simulations can be used to confirm that the selected topologies (it could be interesting to consider more than one topology, in order to select the optimum one afterwards, from the performance factors evaluation) suit the application parameters.

Table 2.1 Design variables and parameters classification used in the thesis

Category	Parameters/Variables
Application parameters	Input voltage range
	Output voltage range
	Output current range
	Output voltage ripple
Static design variables	Inductance value
	Capacitance value
Dynamic design variables	Switching frequency
Performance factors	Energy efficiency
	Occupied area
Technological information	Equivalent parasitic resistances for each component
	Parasitic capacitors for each component
	Capacitive density (capacitance per area)
	Inductive density (inductance per area)

2. Then, the design variables (static and dynamic) impact on the design are determined by means of ideal expressions or model, and some performance factors can be evaluated (i.e. maximum inductor current). At this point, converter topologies options should be shrunk down to a few ones that suit the application; and some design variables range of values should be partially delimited, by checking some application parameters fulfillment (e.g. maximum output voltage ripple). Power switches are considered ideal and no parasitic components or non idealities are considered.

3. The following design step implies a further increase in the model complexity. It is necessary to decide the technological implementation of the different components of the switching power converter. In case of integrated monolithic converters, this requires to select the microelectronic process used to manufacture the design. The foundry provides information about what is feasible or unfeasible to implant in the selected process, and detailed models for components that lead to parasitic characterization. The addition of all this information generates a notable increase in the converter model complexity, and circuit-level simulations become fundamental (although complex system-level simulation are still interesting). The result of this design stage is the evaluation of performance factors such as energy efficiency and volume (or area) occupation, which should lead to an optimized design.

It should be noted that steps 2 and 3 of such design procedure should be carried out for any different selected topology, to choose the one that provides the best performance factors. Furthermore, it is possible that, in spite of the accomplishment of the application parameters, the evaluation of the performance factors results in so poor quality or performance that another topology or design must be considered from the initial point (Fig. 2.1 summarizes the whole process explained above).

Therefore, once the design variables have been identified and their corresponding range of values have been constrained by the application parameters criteria, any combination of design variables values should be explored to evaluate the selected

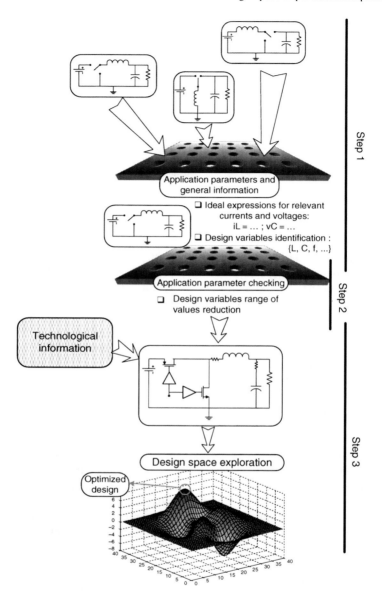

Fig. 2.1 Proposed design procedure. The optimized design is obtained by means of the design space exploration

performance factors. For the sake of simplicity, all the performance factors can be joined in a merit figure definition to obtain a single index that leads to a mathematical maximization, in order to get the desired optimized design. In the following, a representative merit figure definition suitable to mathematical maximization is provided:

$$\Gamma = \frac{A^a B^b}{C^c D^d} \tag{2.1}$$

where A and B are the performance factors to be maximized, whereas C and D should be minimized. The lowercase exponents (a, b, c, and d) are the corresponding weights assigned to each performance factor in the merit figure definition.

The exploration of all the design variables sets of values, searching the optimized design, is what we call the *design space exploration*.

From a mathematical standpoint, the constrained design space exploration (and the corresponding optimization) could be described as follows:

$$minimize \longrightarrow f_o(\mathbf{x}) \tag{2.2}$$

$$subject \ to \longrightarrow f_i(\mathbf{x}) \le b_i \quad i = 1, \cdots, m \tag{2.3}$$

With the following definitions:

$$\mathbf{x} = (x_1, x_2, \cdots, x_n) \longrightarrow design \ variables \ (L, f_s, C_o) \tag{2.4}$$

$$f_o : R^n \to R \longrightarrow objective \ function \tag{2.5}$$

$$f_i : R^n \to R, \quad i = 1, \cdots, m \longrightarrow constraint \ functions \tag{2.6}$$

In general, main function $f_o(\mathbf{x})$ and constraint functions $f_i(\mathbf{x})$ will be non linear (even non continuous functions), which precludes the obtention of the optimized design by means of analytical procedures.

It should be noted that with the inclusion of technological information, the design of each converter component is related to its non idealities, such as its equivalent series resistance (ESR from now on), parasitic capacitances, area occupation. Therefore, given the variables that may appear in each component design, the particular design of the several components that compose the whole converter should be optimized to improve its energy losses or area. In fact, this is one of the strongest points that the microelectronic integration of switching power converters offers: *any component can be particularly optimized for any converter design, (by the designer itself)*, instead of being already implemented by a third-party manufacturer (Fig. 2.2).

As a result of their monolithic integration, one of the performance factors to be evaluated in the design is the *occupied die area*, since it is directly related to the manufacturing cost of the system (in microelectronics, the cost of a chip is proportional to the used area of silicon). The other most important performance factor to be evaluated is the *energy efficiency* in the power conversion, since it is

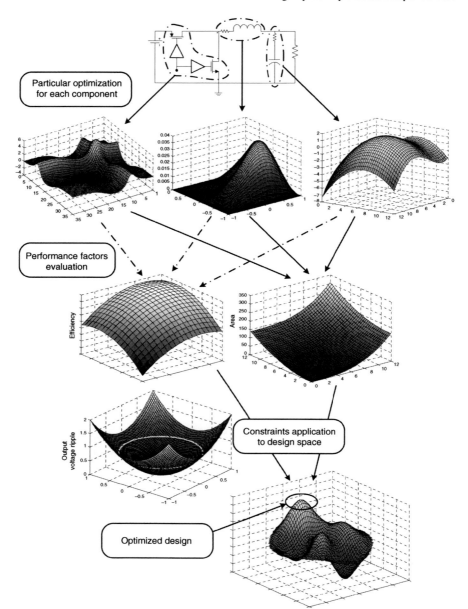

Fig. 2.2 Design space exploration by means of components particular optimizations and the corresponding performance factors evaluation

the main reason in the idea of integrating this kind of power converters and in case of standard CMOS technologies such high efficiency is not assured because those processes have not been conceived for power management applications. A second order performance factor to take into account is the *maximum inductor current*, to reduce the converter components electrical stress. Obviously, whereas energy efficiency should be maximized, occupied silicon area and maximum inductor current are desired to be minimized.

2.2 Simplified Approach for a Buck Converter

The application considered in the development of the design of this work is intended to supply energy to core digital or RF integrated circuits for portable devices, from an external energy source, mainly a Li-Ion battery (that provide a nominal voltage of 3.6 V). In fact, it would become a part of a *Powered-System On-Chip, PSOC*, that integrates all the required functions (or, at least, as many as possible) of the portable device, even the power conditioning system. Because of the continuously decreasing voltage supply of such core circuits in the newest technology nodes (trying to reduce power consumption and their dimensions), as an example, the power converter will be designed to supply a 1 V voltage source. Additionally, the considered nominal output current will be set to 100 mA (which is representative for applications such as power supply of portable devices: MP3 players, some parts of more complex Hand-Held devices, etc...).

Hence, a *conventional Buck converter* is the first topology considered to develop the design, because of its ability of reducing input to output voltage and its simplicity (which can be a very important factor to take into account when attempting to develop a fully integrated prototype).

2.2.1 Continuous and Discontinuous Conduction Modes

Most of the switching power converters can be considered to operate in two different modes depending on the evolution of the energy stored in their corresponding main inductor.

- A switching power converter is understood to work in *continuous conduction mode* (corresponding to the acronymous CCM, from now on) if its inductor current is different from zero at any time, independently of its switching phase (Fig. 2.3a).
- The *discontinuous conduction mode* (DCM from now on) appears in the converter operation if its inductor current becomes and remains zero for a given amount of time, which can range from being almost instantaneous (operation on the edge between DCM and CCM) to last for most of the switching period (Fig. 2.3b).

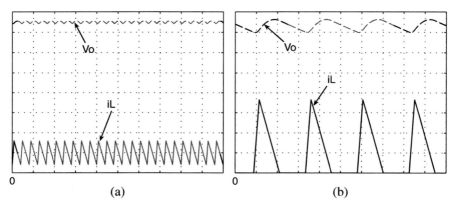

Fig. 2.3 Output voltage and inductor current ideal waveforms for a Buck converter in both operating modes: **a** CCM; **b** DCM

The previous classification becomes of capital importance when analyzing the converter behavior and its corresponding expressions, and, as it will be seen afterwards, the energy losses in each different converter component.

2.2.2 Converter Components Simplified Models

In this section models for all of the converter components are provided. Even though considered as simplified models, these are complex enough to allow the corresponding energy losses calculation as well as its area occupation. Hence, the two first steps of the design procedure proposed in Sect. 2.1 are skipped (since they are covered in detail in the literature), and some technological information is included to allow the performance factors evaluation.

2.2.2.1 Inductor Model

The main inductor of the buck switching power converter is the most difficult component to be integrated on silicon, since in standard CMOS technologies no special materials, such as ferromagnetic cores, are available to increase the self-inductance coefficient.

In power management applications, the main requirements for the inductor design are to provide high inductive density (in order to reduce the occupied silicon area) and very low series resistance (to reduce conduction losses).

However, from the absence of a ferromagnetic core, a benefit is derived because no core magnetization hysteresis cycle appears and thereby no switching losses are present in the inductor. Additionally, no saturation in the magnetic field appears as the inductor current increases (that results in an abrupt equivalent inductance reduction).

Fig. 2.4 Square spiral
inductor

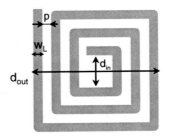

The on-chip implementation by means of standard CMOS technology adds an additional constraint to the inductor design since only planar structures are possible. Therefore, a square-shaped spiral inductor is considered as a first approach design possibility (Fig. 2.4).

In [37] and [38] expressions for inductance value L and total path length l_L (related to its equivalent series resistance, R_L) of a square-shaped inductor are provided as a function of its physical dimensions: number of turns n_L, internal and external diameters (d_{in} and d_{out}), path width w_L and separation between turns p.

$$L = \frac{K_1 \mu n_L^2 d_{avg}}{1 + K_2 \rho} \tag{2.7}$$

where μ is the magnetic permeability of the material ($\mu = 4\pi\,10^{-7}\,H/m$, in absence of ferromagnetic materials) and K_1 and K_2 are constants derived from the data fitting procedure used to find expression (2.7) from a batch of measurements ($K_1 = 2.34$ and $K_2 = 2.75$ for a square inductor). d_{avg} is the average diameter of the spiral and ρ is defined as the *fill ratio*:

$$d_{avg} = 0.5(d_{out} + d_{in}) \tag{2.8}$$

$$d_{in} = d_{out} - 2(n_L w_L + (n_L - 1)p) \tag{2.9}$$

$$\rho = \frac{d_{out} - d_{in}}{d_{out} + d_{in}} \tag{2.10}$$

While the total length of the path is expressed as:

$$l_L = 4 n_L d_{avg} \tag{2.11}$$

Expressions for both inductance and total path length can be obtained as a function of the outer diameter and the number of turns.

$$L = K_1 \mu n_L^2 \frac{\left[d_{out} - (n_L w_L + (n_L - 1)p)\right]^2}{d_{out} + (K_2 - 1)(n_L w_L + (n_L - 1)p)} \tag{2.12}$$

$$l_L = 4 n_L d_{avg} = 4 n_L \left[d_{dout} - (n_L w_L + (n_L - 1)p)\right] \tag{2.13}$$

Assuming that K_1, K_2, μ, w_L and p are technology-dependent constants (this assumption will be further explained afterwards, in another chapter), outer diameter and the number of turns are considered as the inductor design variables.

Because there are many (d_{out}, n_L) pairs that may result in the same value of inductance, total length becomes a function to be minimized (in order to reduce the inductor series resistance), being (d_{out}, n_L) pairs constrained to values that provide the desired L value.

n_L is selected as the independent variable because its values (number of turns of the spiral) are constrained to natural values, being the d_{out} value computed to provide the required inductance. Then, the resulting pairs (d_{out}, n_L) are used to compute the total spiral length. The resulting series resistance will depend on total path length l_L, its width w_L and thickness t_m and the conductivity of the material used to implement the inductor σ.

$$R_L = \frac{l_L}{w_L t_m \sigma} = \frac{l_L R_\square}{w_L} = \frac{4 n_L \left[d_{dout} - (n_L w_L + (n_L - 1)p)\right]}{w_L t_m \sigma} \tag{2.14}$$

In (2.14), the symbol $R_\square$ is the square-resistance of the conductor metal, which is a very common parameter provided by the foundry, in case of planar technologies. It models the resistance provided by an square of conductor, given the conductor thickness (t_m) and conductivity (σ), which are expected to be constant ($R_\square = \frac{1}{t_m \sigma}$).

At first sight, it appears that having assured the desired inductance value, the overall series resistance should be minimized, but an additional constraint should be considered: the total occupied silicon area (A_L), that is a square function of the outer diameter (d_{out}). Therefore, the overall area is computed and a merit figure to be maximized is defined Γ_L, particular for the inductor design itself.

$$A_L = d_{out}^2 = f(L) \tag{2.15}$$

$$\Gamma_L = \frac{1}{A_L R_L} = \frac{w_L t_m \sigma}{d_{out}^2 4 n_L \left[d_{dout} - (n_L w_L + (n_L - 1)p)\right]} \tag{2.16}$$

At this point, it is reminded the dependence $d_{out} = f(L, n_L)$, that leads to the maximization of Γ_L, only as a function of the number of turns. Figure 2.5a shows the evolution of Γ_L. The resulting series resistance and the occupied silicon area for the selected design, are shown as well (Fig. 2.5b).

Having obtained the value of the inductor parasitic series resistance (2.14), the corresponding conduction losses can be determined by means of the inductor *rms* current value.

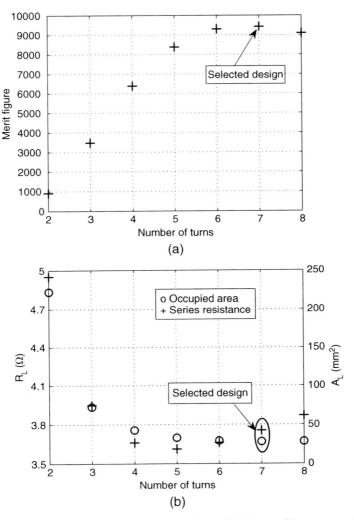

Fig. 2.5 Optimization of an square spiral inductor ($L = 150\,\text{nH}$, $p = 20\,\mu\text{m}$, $w_L = 300\,\mu\text{m}$): **a** merit figure; **b** series resistance and occupied area

$$I_L^2 = \sqrt{\frac{8I_o^3 V_o(V_{bat} - V_o)}{9V_{bat}Lf_s}} \longrightarrow DCM \qquad (2.17)$$

$$I_L^2 = \frac{V_o^2(V_{bat} - V_o)^2 V_{bat}}{12L^2 f_s^2 V_{bat}^3} + I_o^2 \longrightarrow CCM \qquad (2.18)$$

$$P_{L\_cond} = I_L^2 R_L = f(f_s, L) \qquad (2.19)$$

Therefore, as stated in (2.15) and (2.19), it is observed that the occupied area by the inductor and its corresponding power losses are a function of its self-inductance value. Additionally, power losses also depend on the converter switching frequency, being both of them design variables considered in the design space exploration.

2.2.2.2 Capacitor Model

The main requirement for the output capacitor implementation is to offer a high capacitive density, that results in a high capacitance value (to reduce to amount of output voltage ripple) while keeping reduced silicon area occupation. Moreover, to prevent reducing power efficiency and increasing output ripple it is important to get a low equivalent series resistance (ESR) design.

In order to get the highest capacitive density in planar environment, the parasitic gate capacitor of MOS transistors is used to implement the output capacitor (MOS capacitor). The main reason for this choice is that in standard CMOS technologies the gate silicon oxide is the thinnest dielectric material that can be used between both plates of the planar capacitor.

A common given parameter from a CMOS technology is the transistor gate capacitive density (C_{ox}). This is an average parameter since the MOS gate capacitor is a non-linear function of the gate voltage, but it is a good approach if the gate voltage is above the threshold voltage. Therefore, the expression of the area occupied by the MOS capacitor (A_{C_o}) is immediate, from the desired capacitor value:

$$A_{C_o} = \frac{C_o}{C_{ox}} = f(C_o) \tag{2.20}$$

In regard to the capacitor ESR (R_{C_o}), it is a function of the plates and contacts dimensions and materials resistivity. In standard CMOS technologies, the top plate of a MOS capacitor is build up by polysilicon which is quite resistive, and the bottom plate is implemented by the transistor channel, whose conductivity depends on the applied gate voltage. Additionally, it is observed that resistance of both plates is, in fact, a function of the aspect ratio rather than the area. However, in this first approach a constant value could be considered, since the total ESR can be reduced by the parallel connection of several capacitors (keeping the total capacitance value), which is a common technique in case of non-integrated designs.

Finally, from the capacitor current *rms* value (for both operating modes, continuous and discontinuous) the conduction losses associated to the capacitor are directly found:

$$I_{C_o}^2 = (I_L - I_o)^2 = \left(\sqrt[4]{\frac{8I_o^3 V_o(V_{bat} - V_o)}{9V_{bat}Lf_s}} - I_o \right)^2 \longrightarrow DCM \tag{2.21}$$

$$I_{C_o}^2 = (I_L - I_o)^2 = \left(\sqrt{\frac{V_o^2(V_{bat} - V_o)^2 V_{bat}}{12L^2 f_s^2 V_{bat}^3} + I_o^2} - I_o \right)^2 \longrightarrow CCM \quad (2.22)$$

$$P_{C_o} = I_{C_o}^2 R_{C_o} = f(L, f_s) \quad (2.23)$$

In contrast to the inductor design, the direct relation between the capacitor value and its area, as well as the relatively arbitrary and constant ESR value (in this first approach), avoid the need for an optimization procedure of the capacitor design.

Regarding the performance factors dependencies on the design variables, it is observed that the occupied area depends on the capacitance value (2.20), whereas the power losses depend on inductance and switching frequency but not on the capacitance value (provided that the capacitor ESR is independent of this design variable), as stated by (2.23).

2.2.2.3 Power Transistor Model

Power switches suffer from two different power loss mechanisms: *conduction losses* (due to the joule-effect when current flows through its parasitic non-zero channel resistance) and *switching losses* (due to the coexistence of voltage and current excursions during switching instants).

Conduction Losses

In order to derive the conduction losses for each power switch (that in case of CMOS standard technology implementation are MOSFET transistors), the parasitic on-resistance (R_{on}) expression as well as the *rms* value of the current flowing through its channel are required.

$$P_{cond} = R_{on} I_{MOS}^2 \quad (2.24)$$

Current flowing through each transistor is obtained by the corresponding expressions for the case of considering an ideal switching power converter, for the DCM operation case:

$$I_{PMOS\_DCM}^2 = \sqrt{\frac{8I_o^3 V_o^3 (V_{bat} - V_o)}{9L V_{bat}^3 f_s}} \quad (2.25)$$

$$I_{NMOS\_DCM}^2 = \sqrt{\frac{8I_o^3 V_o (V_{bat} - V_o)^3}{9L V_{bat}^3 f_s}} \quad (2.26)$$

And for the CCM:

$$I_{PMOS\_CCM}^2 = \frac{V_o^3(V_{bat} - V_o)^2}{12L^2 V_{bat}^3 f_s^2} + I_o^2 \frac{V_o}{V_{bat}} \tag{2.27}$$

$$I_{NMOS\_CCM}^2 = \frac{V_o^2(V_{bat} - V_o)^3}{12L^2 V_{bat}^3 f_s^2} + I_o^2 \frac{V_{bat} - V_o}{V_{bat}} \tag{2.28}$$

As stated in (2.29), in a first order approach, the on-resistance of a MOS switch depends on the corresponding channel size (L_{ch}, W_{power_MOS} for length and width, respectively), the voltage-to-source voltage (V_{gs}) and technology dependent constants: threshold voltage ($V_{TN/P}$), gate capacitance-per-area (C_{ox}) and carriers mobility ($\mu_{N/P}$). Typically, threshold voltage and carriers mobility will take different values for PMOS and NMOS transistors.

$$R_{on} = \frac{L_{ch}}{\mu_{N/P} C_{ox} W_{power\_MOS}(V_{gs} - V_{TN/P})} \tag{2.29}$$

Since in the case of a buck converter implementation maximum available voltage-to-source corresponds to the input battery voltage (V_{bat}, for a battery operated system) when sources are properly connected to supply voltages, this should be used to drive the MOS switches gate (to minimize the corresponding on-resistance), resulting in the following expression:

$$R_{on} = \frac{L_{min}}{\mu_{N/P} C_{ox} W_{power\_MOS}(V_{bat} - V_{TN/P})} \tag{2.30}$$

being L_{min} the minimum channel length available in a certain technological microelectronic process, to reduce the on-resistance, and the parasitic capacitances as well.

From (2.29) and (2.30) it is derived that conduction losses in a power MOSFET become an inverse function of its channel width.

Switching Losses

Switching losses computation is addressed by means of the product of the voltage across the channel (between transistor drain and source) and the current flowing through it. In order to simplify the model, linear evolutions for both voltage and current are assumed during switching actions, as it can be seen in Fig. 2.6, as well as the corresponding instantaneous power evolution.

In order to obtain the total energy spent in each commutation, instantaneous power has to be integrated along the commutation interval (shaded area in Fig. 2.6).

$$E_{sw} = \frac{I_{on} V_{on} t_{sw\_on}}{6} + \frac{I_{off} V_{off} t_{sw\_off}}{6} \tag{2.31}$$

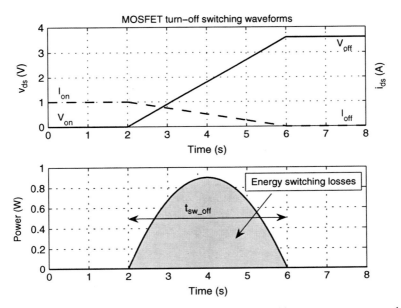

Fig. 2.6 First order approach of channel voltage and current, and instantaneous power evolution along the commutation of a power transistor

Consequently, power switching losses are obtained from energy losses and the switching frequency.

$$P_{sw} = f_s \left(\frac{I_{on} V_{on} t_{sw\_on}}{6} + \frac{I_{off} V_{off} t_{sw\_off}}{6} \right) \tag{2.32}$$

In (2.31) and (2.32), variables I_{on}, V_{on} and t_{sw_on} are the current and voltage swings applied to the MOS transistor when turning it on, and the corresponding duration; whereas the second terms of those expressions are referred to the turning off action.

Table 2.2 collects the expressions corresponding to the voltage and current conditions applied to both power transistors, for DCM and CCM.

While V_{on} and I_{on} depend on the converter design and operation, the $t_{sw\_on/off}$ values depend on the transistor and driver design. As a first order approach, state transition duration of a MOS power switch can be estimated assuming that its gate voltage evolution corresponds to the gate capacitance charge/discharge process through the equivalent resistance of the driver last stage. In this case, a 3-time-constants criterion can be used to determine the duration of the total change in the gate voltage.

$$t_{sw} = 3 R_{driver} C_{gate} = 3 R_{driver} C_{ox} W_{power\_MOS} L_{min} \tag{2.33}$$

Table 2.2 Transistor switching conditions for DCM and CCM operation

DCM	PMOS	NMOS
V_{on}	$V_{bat} - V_o$	V_{bat}
I_{on}	0	$I_{Lmax} = \sqrt{\dfrac{2I_o V_o (V_{bat} - V_o)}{L f_s V_{bat}}}$
V_{off}	V_{bat}	V_o
I_{off}	$I_{Lmax} = \sqrt{\dfrac{2I_o V_o (V_{bat} - V_o)}{L f_s V_{bat}}}$	0
CCM	**PMOS**	**NMOS**
V_{on}	V_{bat}	V_{bat}
I_{on}	$I_{Lmin} = I_o - \dfrac{V_o (V_{bat} - V_o)}{2 L V_{bat} f_s}$	$I_{Lmax} = I_o + \dfrac{V_o (V_{bat} - V_o)}{2 L V_{bat} f_s}$
V_{off}	V_{bat}	V_{bat}
I_{off}	$I_{Lmax} = I_o + \dfrac{V_o (V_{bat} - V_o)}{2 L V_{bat} f_s}$	$I_{Lmin} = I_o - \dfrac{V_o (V_{bat} - V_o)}{2 L V_{bat} f_s}$

Hence, at this point of the analysis, it is observed from (2.32) and (2.33) that switching losses become a direct function of the MOS transistor channel width. The value of equivalent driver output resistance will be derived in the following Sect. 2.2.2.4, which deals with the driver design issues.

To finish the power switch model, their corresponding occupied silicon area is computed as a function of their gate dimensions.

$$A_{MOS} = W_{power\_MOS} L_{min} \tag{2.34}$$

2.2.2.4 Power Driver Design and Loss Model

A power driver is required to amplify the power of the switching signal when it has to drive the power MOSFET gate input capacitance (which usually is rather large), if a fast gate state transition is desired. In standard CMOS technology implementations the most usual way to implement a bi-state power driver is by means of a tapered buffer (Fig. 2.7), which consists of a chain of digital inverters; being the transistors of each stage larger than those from the previous by a tapering factor f.

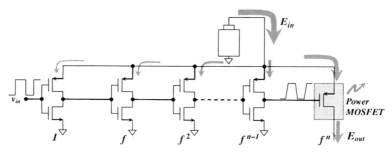

Fig. 2.7 Power driver implementation by means of tapered buffers

Given the size of the transistor of a minimum digital inverter (or its equivalent input capacitance), that should be the first of the tapered buffer, and the size of the power MOSFET to be driven (or its equivalent input capacitance), two different variables are available to design the tapered buffer: the number of inverters n and the tapering factor. Usually, tapered buffer designs with a constant tapering factor are implemented, and then the relationship between the tapering factor and the number of stages is as follows:

$$f = \sqrt[n]{\frac{C_{out}}{C_{in}}} \tag{2.35}$$

where C_{out} is the power MOSFET gate capacitance and C_{in} is the minimum inverter capacitance.

Typically, in digital microelectronics research fields, the tapering factor has been determined targeting the overall propagation delay minimization, and it has been found the number e to be the optimum one. Having determined the optimum tapering factor, the number of stages is directly derived from the input-to-output capacitance ratio.

$$n = \ln \left(\frac{W_{power\_MOS}}{(1 + w_{pn}) W_{min}} \right) \tag{2.36}$$

being W_{min} the channel width of the NMOS transistor of the first inverter, and w_{pn} the relationship between the channel width of PMOS and NMOS transistor of each inverter. In (2.36) it is supposed that transistors from inverters, as well as the power MOS, have minimum channel length, to increase switching speed and reduce parasitic resistance.

Having determined the tapering factor and the number of stages of the whole tapered buffer, the equivalent output resistance that charges and discharges the power MOS gate capacitance can be obtained (R_{driver}), through the channel width of the NMOS transistor of the last driver stage (W_{n_driver}).

$$W_{n\_driver} = W_{min} e^{n-1} = \frac{W_{power\_MOS}}{(1 + w_{pn}) e} \tag{2.37}$$

$$R_{driver} = \frac{L_{min}}{\mu_N C_{ox} W_{n\_driver} (V_{bat} - V_{TN})} =$$
$$= \frac{(1 + w_{pn}) e L_{min}}{\mu_N C_{ox} W_{power\_MOS} (V_{bat} - V_{TN})} \tag{2.38}$$

In Addition to the driver output resistance, its own losses must be modeled in order to include them into the design space exploration, as a part of the overall efficiency.

In case of a tapered buffer, power losses are mainly due to its state change at every switching period. As a first order approach, power losses are computed by means of the charge spent from the power supply to change the value of the respective input gate capacitances of all inverters of the chain.

As a consequence, the total gate capacitance (C_T) should be computed as a function of the number of stages and the tapering factor:

$$C_T = \sum_{i=0}^{n} C_i = (1 + w_{pn}) C_{ox} W_{min} L_{min} \sum_{i=0}^{n} f^i \qquad (2.39)$$

$$\sum_{i=0}^{n} f^i = \frac{f^{n+1} - 1}{f - 1} \qquad (2.40)$$

being e the optimum tapering factor:

$$C_T = (1 + w_{pn}) C_{ox} W_{min} L_{min} \frac{e^{n+1} - 1}{e - 1} \qquad (2.41)$$

And the corresponding power losses are related to the power supply voltage and the switching frequency by:

$$P_{driver} = V_{bat}^2 f_s (1 + w_{pn}) C_{ox} W_{min} L_{min} \frac{e^{n+1} - 1}{e - 1} \qquad (2.42)$$

From (2.42) it is observed the driver switching losses increase with the number of inverters that, in turn, is an increasing function of the power transistor channel width. Thus, it is concluded that power driver losses are a linear function of the power MOSFET channel width.

$$P_{driver} = V_{bat}^2 f_s (1 + w_{pn}) C_{ox} W_{min} L_{min} \frac{e^{\frac{W_{power\_MOS}}{(1+w_{pn})W_{min}}} - 1}{e - 1} \qquad (2.43)$$

$$P_{driver} = V_{bat}^2 f_s C_{ox} L_{min} \frac{e W_{power\_MOS} - (1 + w_{pn}) W_{min}}{e - 1} \qquad (2.44)$$

On the other hand, the overall occupied silicon area can be computed directly from the total gate capacitance of the tapered buffer if only channel dimensions are considered.

$$A_{driver} = L_{min} \frac{e W_{power\_MOS} - (1 + w_{pn}) W_{min}}{e - 1} \qquad (2.45)$$

2.2.2.5 Power MOSFET and Driver Joint Design

In previous Sects. 2.2.2.3 and 2.2.2.4 power losses were derived as a function of the power MOSFET channel width. Conduction losses of MOS switch were found to be a decreasing of the channel width, whereas switching losses depend not only on the power switch channel width but on the driver output resistance.

Since output driver resistance was determined in expression (2.38), a closed analytical expression can be found for power MOSFET switching losses, by means of the switching action duration definition.

In the case in which the equivalent driver output resistance for both charge and discharge output capacitance were the same (that can be achieved if a proper relationship between PMOS and NMOS transistor channel width is selected, for the tapered buffer inverters), the duration on the two complementary actions on the power switch would be the same.

$$t_{sw\_on/off} = \frac{3(1 + w_{pn})eL_{min}^2}{\mu_N(V_{bat} - V_{TN})} \qquad (2.46)$$

Which results in the following switching power losses expression:

$$P_{sw} = \frac{(1 + w_{pn})eL_{min}^2}{2\mu_N(V_{bat} - V_{TN})}(V_{on}I_{on} + V_{off}I_{off})f_s \qquad (2.47)$$

From (2.47) it is observed that *switching power losses from the power MOSFET become independent of its channel width*, as long as the corresponding driver design becomes adapted to it.

And finally, from (2.44) it is observed that driver power losses are a direct linear function of the power MOSFET channel width.

In Table 2.3, the expressions of the three different power loss mechanisms related to the power switches and their corresponding trends as W_{power_MOS} increase are summarized.

Since there are two opposed trends, an optimum design that provides minimum power losses is expected as a function of the power transistor channel width, as shown in the following figure (corresponding to a PMOS design and its associated driver) (Fig. 2.8).

In the following, expressions for the optimum channel width value, and the resulting minimum power losses are provided:

Table 2.3 Power loss expressions and trends for the MOSFET and driver designs

Loss source	Expression	Trend (as W_{power_MOS} increases)
Power MOS conduction losses	$P_{cond} = \dfrac{L_{min}}{\mu_{N/P}C_{ox}W_{power\_MOS}(V_{bat} - V_{TN/P})}I_{MOS}^2$	↓↓
Power MOS switching losses	$P_{sw} = \dfrac{(1 + w_{pn})eL_{min}^2}{2\mu_N(V_{bat} - V_{TN})}(V_{on}I_{on} + V_{off}I_{off})f_s$	Constant
Driver losses	$P_{driver} = V_{bat}^2 f_s C_{ox} L_{min} \dfrac{eW_{power\_MOS} - (1 + w_{pn})W_{min}}{e - 1}$	↑↑

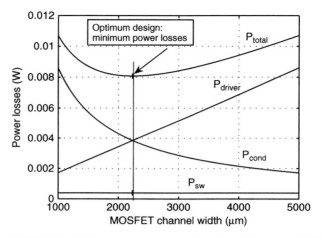

Fig. 2.8 Power MOSFET and driver power losses as a function of the power MOSFET channel width

$$W_{opt} = \frac{I_{MOS}}{V_{bat} C_{ox}} \sqrt{\frac{e - 1}{\mu_{N/P} f_s e (V_{bat} - V_{TN/P})}} \quad (2.48)$$

$$P_{MOS+driver} = f_s (1 + w_{pn}) L_{min}$$
$$\left(\frac{e L_{min}}{2 \mu_N (V_{bat} - V_{TN})} (V_{on} I_{on} + V_{off} I_{off}) - C_{ox} V_{bat}^2 \frac{W_{min}}{e - 1} \right) +$$
$$+ 2 I_{MOS} V_{bat} L_{min} \sqrt{\frac{f_s e}{\mu_{N/P} (e - 1)(V_{bat} - V_{TN/P})}} \quad (2.49)$$

2.2.3 Output Voltage Ripple

Once the converter topology has been selected to fulfill some of the application parameters herein considered (mainly, input voltage to output voltage conversion ratio), there is still an application parameter to take into account to yield the desired design. This is the *output voltage ripple*, that becomes a hard constraint of the design space, since any design variables set of values (L, f_s, C_o) that results in an output voltage ripple greater than the one tolerated by the load circuitry to be supplied, will be neglected.

Typically, the output voltage ripple Δv_o is defined as the difference between the maximum and minimum values that the output voltage takes throughout a whole switching period.

$$\Delta v_o = V_{o\_max} - V_{o\_min} \quad (2.50)$$

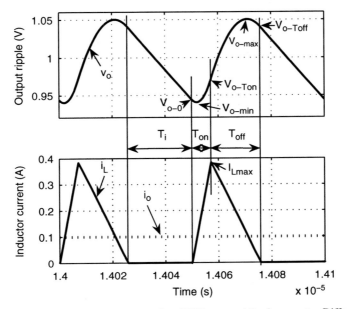

Fig. 2.9 Output voltage and inductor current in a DCM operated Buck converter. Different phases of a switching period are detailed

Figures 2.9 and 2.10 explicitly show these values for DCM and CCM operation. Additionally, other significant and useful characteristics of the output voltage and inductor current waveforms for both operating modes are also noted.

The corresponding expressions for V_{o_max} and V_{o_min} are obtained from the low output ripple assumption, which implies that inductor current correspond to a piecewise linear function. According to the scheme from Fig. 2.11, the output voltage generic expression for both operating modes and different phases that compose a whole switching cycle, is the following.

$$v_o = i_{C_o} R_{C_o} + \frac{1}{C_o} \int i_{C_o} dt = (i_L - I_o) R_{C_o} + \frac{1}{C_o} \int i_L dt - \frac{I_o}{C_o} + V_{C_o\_0}$$

$$(2.51)$$

where $V_{C_o\_0}$ is the output capacitor voltage (without ESR) at the beginning of the T_{on} phase.

Taking into account different i_L expressions for any operating mode and switching phase, (2.2.3) is minimized or maximized accordingly to minimum or maximum research.

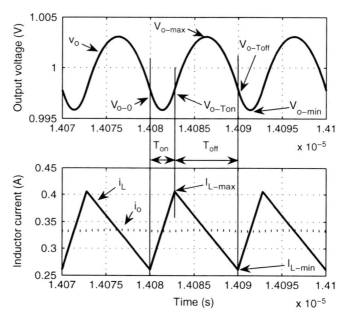

Fig. 2.10 Output voltage and inductor current in a CCM operated Buck converter. Different phases of a switching period are detailed

Fig. 2.11 Buck converter output circuit

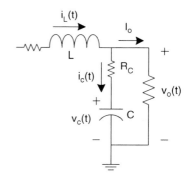

In case of DCM, the expressions for V_{o_max} and V_{o_min} are:

$$V_{o\_max} = \frac{[(I_{L\_max} - I_o)L - R_{C_o}C_oV_o]^2}{2LC_oV_o} + \frac{V_{bat} - V_o}{2LC_o}T_{on}^2 +$$

$$+ \frac{(V_{bat} - V_o)R_{C_o}C_o - I_oL}{LC_o}T_{on} - I_oR_{C_o} + V_{C_o\_0} \qquad (2.52)$$

$$V_{o\_min} = V_{C_o\_0} - I_oR_{C_o} - \frac{[I_oL - R_{C_o}C_o(V_{bat} - V_o)]^2}{2LC_o(V_{bat} - V_o)} \qquad (2.53)$$

But the following constraints must be considered:

$$RC_o > \frac{(I_{L\_max} - I_o)L}{V_o C_o} \longrightarrow V_{o\_max} = v_o(T_{on}) = \frac{V_{bat} - V_o}{2LC_o}T_{on}^2 +$$

$$+ \frac{(V_{bat} - V_o)RC_o C_o - I_o L}{LC_o}T_{on} - I_o RC_o + V_{C_o\_0} \tag{2.54}$$

$$RC_o > \frac{I_o L}{(V_{bat} - V_o)C_o} \longrightarrow V_{o\_min} = v_o(0) = V_{C_o\_0} - I_o RC_o \tag{2.55}$$

In expressions from (2.52), (2.53), (2.54) and (2.55) T_{on} and I_{L_max} correspond to the T_{on} phase duration and the maximum inductor current (see Fig. 2.9), respectively (for DCM operation).

$$I_{L\_max} = \sqrt{\frac{2I_o V_o L}{V_{bat} f_s (V_{bat} - V_o)}} \tag{2.56}$$

$$T_{on} = \sqrt{\frac{2I_o V_o (V_{bat} - V_o)}{V_{bat} f_s L}} \tag{2.57}$$

When the CCM operation is considered the expressions for the boundary values of output ripple are as follows:

$$V_{o\_max} = \frac{V_o[(V_{bat} - V_o)^2 + (2V_{bat} RC_o C_o f_s)^2]}{8LC_o V_{bat}^2 f_s^2} + V_{C_o\_0} \tag{2.58}$$

$$V_{o\_min} = V_{C_o\_0} - \frac{V_{bat} - V_o}{2LC_o}\left[R_{C_o}^2 C_o^2 + \left(\frac{V_o}{2V_{bat} f_s}\right)^2\right] \tag{2.59}$$

And the corresponding constraints to be taken into account are:

$$RC_o > \frac{V_{bat} - V_o}{2V_{bat} C_o f_s} \longrightarrow V_{o\_max} = v_o(T_{on}) = \frac{(V_{bat} - V_o)V_o RC_o}{2Lf_s V_{bat}} + V_{C_o\_0} \tag{2.60}$$

$$RC_o > \frac{V_o}{2V_{bat} C_o f_s} \longrightarrow V_{o\_min} = v_o(0) = V_{C_o\_0} - \frac{(V_{bat} - V_o)V_o}{2LV_{bat} f_s}RC_o \tag{2.61}$$

All the V_{o_max} and V_{o_min} previous expressions, contain the initial voltage of output capacitor ($V_{C_o\_0}$), which is directly related to the average converter output voltage. It is important to note that for ripple calculation purposes, this value is not necessary because just the difference between both values is required.

Moreover, in both cases (DCM and CCM), the obtained expressions are subject to some constraints depending on the RC_o value. This results from the fact that the minimization or maximization of (2.2.3) is limited to T_{on} and T_{off} phases, respectively. Nevertheless, under some circumstances (stated by inequalities related to RC_o), the minimum or maximum of such parabolic sections, could be somewhere outside those boundaries. In these cases, minimum and maximum output voltage becomes the output voltage at the beginning of the corresponding switching phase.

In summary, this is the reason that precludes the existence of a closed analytical expression for the output voltage ripple.

Figure 2.12 shows good matching between the proposed expressions for output ripple calculation (as a function of the capacitor resistance) and some simulation results, for both operating modes: DCM ($V_o = 1\,\mathrm{V}$, $V_{bat} = 3.6\,\mathrm{V}$,

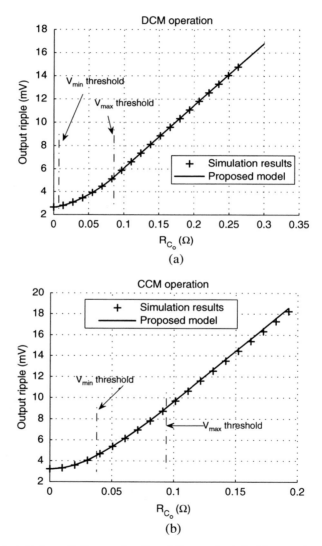

Fig. 2.12 Output ripple prediction as a function of R_{C_o}, contrasted with some simulation results, for DCM **a** and CCM **b**

$L = 50\,\text{nH}$, $C_o = 25\,\text{nF}$, $f_s = 100\,\text{MHz}$, $I_o = 10\,\text{mA}$) and CCM ($V_o = 1\,\text{V}$, $V_{bat} = 3.6\,\text{V}$, $L = 50\,\text{nH}$, $C_o = 25\,\text{nF}$, $f_s = 150\,\text{MHz}$, $I_o = 100\,\text{mA}$). In this figure, threshold values for R_{C_o} exposing the constraints from (2.54), (2.55), (2.60) and (2.61) are marked with vertical lines. It is interesting to note that these constraints (and the corresponding change in calculation expressions) appear for low R_{C_o} values.

However, it should be taken into account that as a the load resistance value becomes lower (and closer to the R_{C_o}), little error occurs on ripple prediction.

2.2.4 Design Space Exploration Results

In the previous sections, all the converter components have been modeled towards the selected performance factors evaluation (power efficiency and occupied area) and application parameters (output voltage ripple). At this point, with the information provided from such models, the design space exploration is carried on by means of the power efficiency evaluation.

$$\eta(\%) = \frac{P_{out}}{P_{out} + P_{losses}} 100 =$$

$$= \frac{V_o I_o}{V_o I_o + P_{NMOS+driv} + P_{PMOS+driv} + P_{C_o} + P_{L\_cond}} 100 \quad (2.62)$$

In Table 2.4, all the required technological information to evaluate the design performance factors, can be found (which correspond to a representative standard CMOS 0.25 μm process).

Expression (2.62) generates the results depicted in Fig. 2.13, which expose 3 different 3D surfaces since the function to be represented is a fourth dimensional one (efficiency plus the 3 design variables). Therefore, efficiency as a function of L

Table 2.4 Technological parameters information required to develope the design space exploration

Component	Technological parameter	Value
Inductor	$R_\square$	$13\,\text{m}\Omega/\square$
	μ	$4\pi\,10^{-7}\,\text{H/m}$
	w_L	$300\,\mu\text{m}$
	p	$20\,\mu\text{m}$
Capacitor	R_{C_o}	$50\,\text{m}\Omega$
	C_{ox}	$4.933\,\text{fF}/\mu\text{m}^2$
Power MOSFETs	W_{min}	$0.3\,\mu\text{m}$
	L_{min}	$0.34\,\mu\text{m}$
	μ_n	$393.8 \times 10^8\,\mu\text{m}^2/(\text{sV})$
	μ_p	$89 \times 10^8\,\mu\text{m}^2/(\text{sV})$
	V_{TP}	$0.65\,\text{V}$
	V_{TN}	$0.55\,\text{V}$

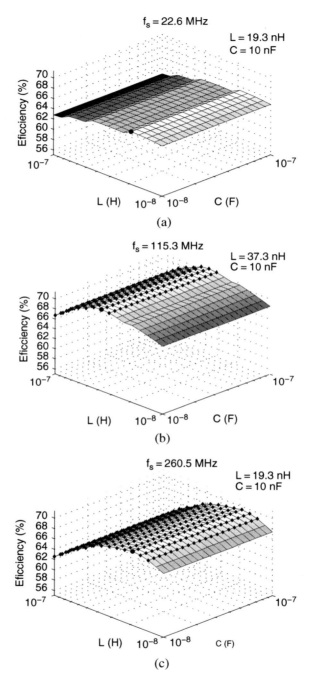

Fig. 2.13 Buck converter power efficiency: **a** $f_s = 22.6\,\text{MHz}$; **b** $f_s = 115.3\,\text{MHz}$; **c** $f_s = 260.5\,\text{MHz}$

and C_o is depicted, and all figures correspond to a switching frequency single value. In the center figure the maximum power efficiency is achieved.

In Fig. 2.13 a black dot is depicted to mark, in the (L, C_o) plane, the design that provides the maximum efficiency, whereas black diamonds are used to mark all the designs that will result in CCM operation for the considered output current (100 mA).

As expected, it is observed that power efficiency is a function of inductor and switching frequency values, whereas it becomes constant against changes in output capacitor value, since no relationship between C_o and R_{C_o} have been considered in this simplified case (despite just one dot is depicted, any row with C_o constant results in the same power efficiency).

In fact, due to this particularity power efficiency is represented as a function of f_s and L in Fig. 2.14, where the maximum efficiency is better observed.

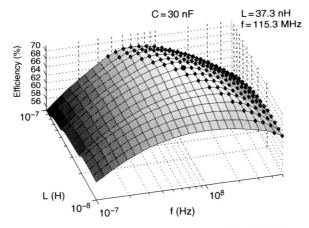

Fig. 2.14 Buck converter power efficiency in the (f_s, L) plane $(C_o = 30 \, \text{nF})$

Since conduction mode is independent of the output capacitor (if low output ripple is assumed), as the power efficiency, Fig. 2.14 shows all the design solutions that will produce CCM operation (marked with black diamonds).

As far as occupied area (2.64) is concerned, the same kind of figures are presented in Fig. 2.15. It is noted that the area variation with f_s is insignificant. The reason for this is that the total converter area depends most on the inductor and capacitor values (which are independent of switching frequency). Only the power MOSFETs and their corresponding drivers design are frequency dependent, but its impact on the overall area is negligible. Additionally, in the 3 depicted cases, it is clearly observed how, as the switching frequency is increased, more designs become CCM operated.

Following the obtention of both considered performance factors (total area occupied and power efficiency), the merit figure to be maximized can now be defined (equivalent to $f_o(x)$, exposed in (2.2)).

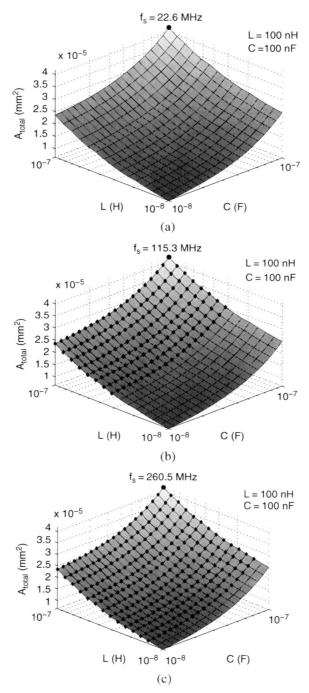

Fig. 2.15 Buck converter occupied area: **a** $f_s = 22.6\,\text{MHz}$; **b** $f_s = 115.3\,\text{MHz}$; **c** $f_s = 260.5\,\text{MHz}$

$$\Gamma = \frac{\eta(\%) - \eta_{min}(\%)}{A_{total}} \qquad (2.63)$$

$$A_{total} = A_L + A_{C_o} + A_{PMOS} + A_{NMOS} + A_{driverN} + A_{driverP} \qquad (2.64)$$

In (2.63), the minimum obtained power efficiency is substracted from the rest of the power efficiency values as a way to increase the effect of its span, because of its relatively narrow range (from 54.5 to 71.4%).

Figure 2.16 is a volumetric representation of the (2.63) results, where the darker the painted area is, the higher the merit figure value becomes. It is just intended to be a qualitative representation revealing the existence of a single maximum (corresponding to an optimized design), and, approximately, where it is obtained (in terms of the coordinates (L, C_o, f_s)). Due to difficulties of volumetric representation, several slices are depicted on all axis, as well as a 3-dimensional contour around the maximum value.

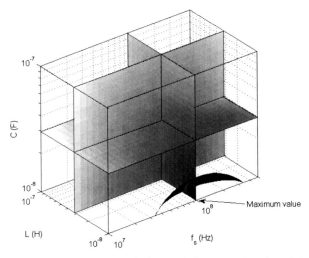

Fig. 2.16 Volumetric representation of the design merit figure as a function of the 3 design variables (L, C_o, f_s). In the figure, the *darker area* is, the higher the merit figure value becomes

Figure 2.17 shows 3 different surfaces where the merit figure is exposed as a function of inductor and capacitor values for 3 different f_s values, respectively (the picture in the middle corresponds to the optimum switching frequency $f_s = 98\,\text{MHz}$).

As observed, the maximum merit figure value is obtained when minimum inductor and capacitor are applied, because this greatly reduces the occupied area, whereas the optimum switching frequency is the one that increases power efficiency. Table 2.5 summarizes the main characteristics of this optimized design.

Apparently, an optimized design (in terms of performance factors) has been obtained, but there is still a constraint to be taken into account. It could be possible that the obtained design results in an output ripple higher than the maximum

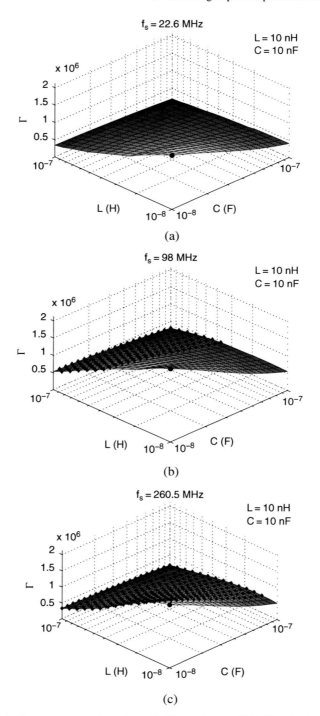

Fig. 2.17 Merit figure representation, from (2.63): **a** $f_s = 22.6\,\text{MHz}$; **b** $f_s = 98\,\text{MHz}$; **c** $f_s = 260.5\,\text{MHz}$

Table 2.5 Main characteristics of the optimized design before the application of the output ripple constraint

Variable	Value
Battery voltage	3.6 V
Output voltage	1 V
Output current	100 mA
Inductor (L)	10 nH
Capacitor (C_o)	10 nF
Switching frequency (f_s)	98 MHz
Total Area	6.59 mm^2
Power efficiency	68.6%
Output voltage ripple	60.2 mV

allowed by the considered application, specially if minimum inductor and capacitor are implemented.

Consequently, the following step is to evaluate the output voltage ripple, and observe how the design space is constrained by those designs that result in an output ripple higher than the maximum allowed by the load circuitry (in this work, the maximum allowed is 50 mV).

Thus, using expressions from Sect. 2.2.3 the output voltage ripple produced by any considered design is calculated and represented in Fig. 2.18. Again, 3 different figures are shown, corresponding to the 3 different switching frequency single values (22.6 MHz, 98 MHz, 260.5 MHz). In this case, it is also marked the maximum allowed ripple by means of a contour line (black solid line). Additionally, a black dot is used to mark the design that results in maximum ripple for each switching frequency value, as well as black diamonds are used to mark the CCM operated designs.

As it would be expected, maximum output ripple is produced when minimum L and C_o values are implemented. It is important to note that in case of $f_s = 98$ MHz output ripple is almost not a constraint of the (L, C_o) plane, and furthermore, for $f_s = 260.5$ MHz (Fig. 2.18c), it is no more a constraint (obviously, the overall design space is already constrained by the high switching frequency values).

In Fig. 2.19, a 2-dimensional view of the design space reduction is shown by means of a family of isolines (corresponding to different f_s values), that shrink down the possible designs area as the switching frequency becomes lower.

Applying the output voltage ripple constraint, power efficiency is depicted again in Fig. 2.20, as well as the total area in Fig. 2.21.

As expected from Fig. 2.19 the (L, C_o) design space is strongly constraint for low switching frequencies (e.g. $f_s = 22.6$ MHz), while it becomes in a minor reduction for $f_s = 98$ MHz or even unchanged at higher switching frequency.

Nevertheless, the output ripple constraint modifies the localization of the optimized design: Fig. 2.22 repeats the merit figure results exposed in Fig. 2.17, but this time the design space have been reduced, and the previous optimized design is not feasible since it produces an unacceptable output ripple. Please note that in Fig. 2.22, axis orientation has been rotated by 180° to increase the constraint effect visibility.

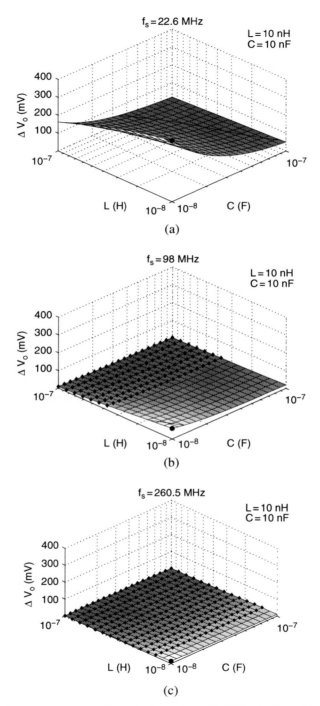

Fig. 2.18 Buck converter output voltage ripple: **a** $f_s = 22.6\,\text{MHz}$; **b** $f_s = 98\,\text{MHz}$; **c** $f_s = 260.5\,\text{MHz}$

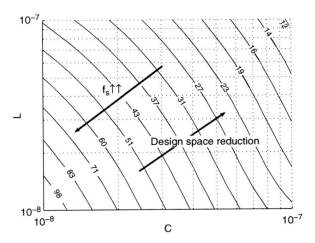

Fig. 2.19 Design space reduction (in plane (L, C_o)) due to the output voltage ripple constraint, as a funtion of the switching frequency (numbers on isolines represent f_s values in MHz). Possible designs appear on the *right* of any isoline

Hence, the optimized design previously found in (Fig. 2.17b) is replaced by the one that appears in Fig. 2.23. It should be noted that although the inductor and capacitor values remain the same (which results in the minimum area design), switching frequency is increased in order to produce lower output ripple. Obviously, the resulting power efficiency is slightly lower than the one corresponding to (Fig. 2.17b). Table 2.6 summarizes the complete set of characteristics of this design setup. In this case, very small change on efficiency results from the frequency increase (which in turn is due to the output ripple constraint application), although in other cases design space constraint could have a greater effect on the final result.

In Fig. 2.24 power loss distribution as well as occupied area are shown as a function of the power lower loss mechanisms and components, respectively.

As regards energy losses distribution, it is found what would have been expected from results in Sect. 2.2.2.5:

- Transistor switching losses are independent not only from the transistor channel width, but even from the transistor type (NMOS or PMOS), since their corresponding driver will be accordingly designed.
- In both types of transistors, those associated energy losses that depend on channel width (conduction and driver losses) become equalized when losses are minimized. This is caused by the opposed trends that these loss mechanisms present when channel width is changed.
- Inductor conduction losses are the highest, which is reasonable because any current flowing through the Buck converter always passes through the inductor, whereas it is split between both transistors.
- Capacitor conduction losses are the lowest because of the low capacitor ESR considered, and the fact that its current is much lower than the inductor current.

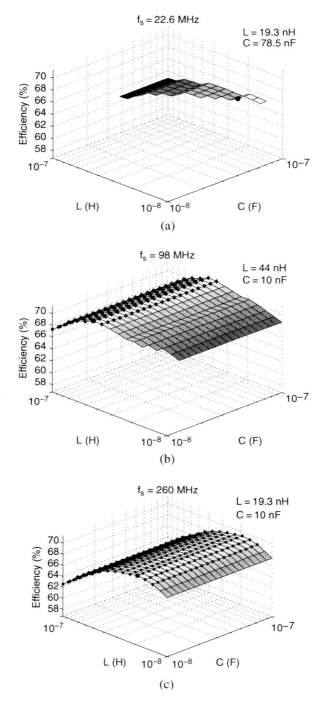

Fig. 2.20 Buck converter power efficiency after applying the output ripple constraint on design space: **a** $f_s = 22.6\,\text{MHz}$; **b** $f_s = 98\,\text{MHz}$; **c** $f_s = 260.5\,\text{MHz}$

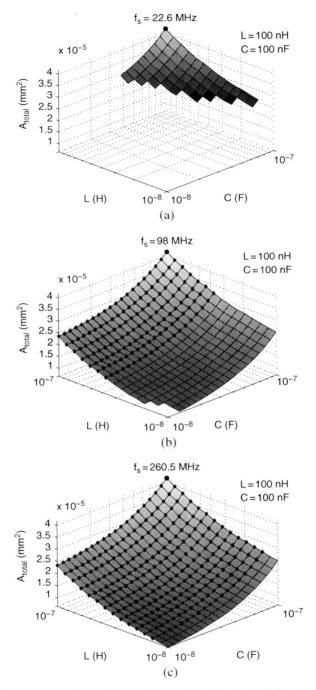

Fig. 2.21 Buck converter total occupied area after applying the output ripple constraint on design space: **a** $f_s = 22.6$ MHz; **b** $f_s = 98$ MHz; **c** $f_s = 260.5$ MHz

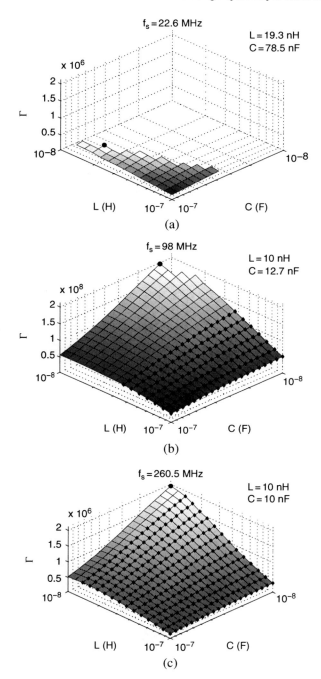

Fig. 2.22 Merit figure representation, applying the output ripple constraint: **a** $f_s = 22.6$ MHz; **b** $f_s = 98$ MHz; **c** $f_s = 260.5$ MHz

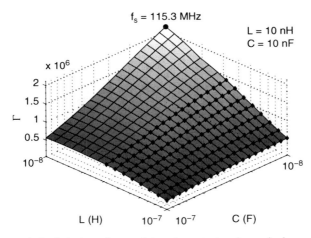

Fig. 2.23 New optimized design after applying the output voltage ripple constraint (f_s = 115.3 MHz)

Table 2.6 Complete set of characteristics of the optimized design after the application of the output ripple constraint

Variable	Value
Battery voltage	3.6 V
Output voltage	1 V
Output current	100 mA
Inductor (L)	10 nH
Inductor ESR	621.5 mΩ
Inductor outer diameter	2.1 mm
Inductor number of turns	3
Capacitor (C_o)	10 nF
Capacitor ESR	50 mΩ
Switching frequency (f_s)	115.3 MHz
Total Area	6.59 mm²
Power efficiency	68.57%
Output voltage ripple	49 mV
Maximum inductor current	354 mA
Operating mode	DCM
N-MOSFET channel width	1,572 μm
N-MOSFET driver inverters	7
P-MOSFET channel width	2,083 μm
P-MOSFET driver inverters	7

In addition to this, Fig. 2.24b clarifies that the occupied silicon area is almost absolutely due to the energy storage components.

The main conclusion that arises from the presented design procedure is that the design challenge that any designer faces when trying to develop a switching power converter to fulfill some specifications is solved by means of the design space exploration. This just requires the appropriate modelling of any of the components to be designed, and subsequently the application of the considered constraints of the

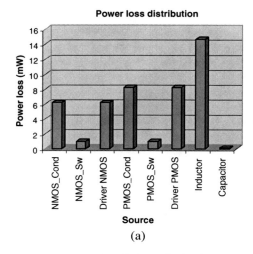

Power loss distribution

Source	Power losses (mW)
NMOS_Cond	6,23
NMOS_Sw	1,02
Driver NMOS	6,23
PMOS_Cond	8,26
PMOS_Sw	1,02
Driver PMOS	8,26
Inductor	14,66
Capacitor	0,14

(a)

Occupied area distribution

Component	Area (mm²)
NMOS	$5,3 \cdot 10^{-4}$
Driver NMOS	$3,1 \cdot 10^{-4}$
PMOS	$7,1 \cdot 10^{-4}$
Driver PMOS	$4,1 \cdot 10^{-4}$
Inductor	4,56
Capacitor	2,03

(b)

Fig. 2.24 Optimized design power losses **a** and occupied area **b** distributions, for each converter component

design space. Finally, the convenient definition of a merit figure is used to obtain the desired converter configuration.

Therefore, it is observed that the huge number of questions that involve the whole design (reactive components values, switching frequency, power transistors size, power drivers design, ...) are indirectly answered, being the final result a set of design characteristics and values as the exposed in Table 2.6, after considering all the suitable possibilities.

Chapter 3
Contributions on Converter Integrated Components and Detailed Models

Abstract In this chapter, proposals for the integrated implementation of some of the converter components (this is, the inductor and the capacitor) are presented. Additionally, detailed models of all the components, mainly focused on the power losses and occupied area evaluation are also proposed. A triangular spiral composed by bonding wires is proposed to implement the inductor, and its design optimization is provided as well. With respect to the output capacitor implementation, a bi-dimensional matrix of parallel connected MOS capacitors (MOSCAP) is proposed. In addition to the design optimization of a single MOSCAP (to minimize its ESR), a model for the whole structure ESR is presented, as well as its optimization. Regarding the tapered buffers design (use to implement the power drivers), an energy consumption model is proposed, as well as the model for their output fall-rise time, which is used as a link required for the concurrent design of the power drivers and the power transistors. Finally, focusing on the power transistors, a model for the switching losses is presented (from a revised existing model), which splits the switching losses into two different mechanisms, the capacitive switching losses and the resistive switching losses. At the end of the section, the complete procedure for the concurrent optimization of both the power transistors and their corresponding drivers is proposed.

3.1 Bonding Wire Inductor Model

3.1.1 State of the Art

The traditional 3-dimensional conception of inductive structures has become the main handicap for their monolithic integration, since in standard CMOS ICs only planar designs are viable. However, their great interest and usefulness in circuit design have pushed many researchers to work in their integration on silicon via different approaches. Roughly, these can be classified in two different research lines: those that use MEMS techniques and their capabilities, and those that remain constrained by the possibilities of standard CMOS processes.

G. Villar Piqué, E. Alarcón, *CMOS Integrated Switching Power Converters*,
DOI 10.1007/978-1-4419-8843-0_3, © Springer Science+Business Media, LLC 2011

In front of this and in the context of this thesis, the main requirements for inductors designed targeting power converters should be considered. Since the main inductor of a switching power converter is considered as an energy storage component, its main required characteristics are:

- Low equivalent series resistance (ESR), to reduce the energy losses due to Joule-Effect when electrical current flows through the inductor.
- High inductance coefficient, which means that high amounts of energy can be stored in the corresponding magnetic field, without the need of high current values. Focusing on the implementation of an integrated design, in this work component values are often considered relative to area occupation. Therefore, a high inductive density (ratio between inductance value and occupied area) is desired.
- If a ferromagnetic core is used to increase the inductance value (and also in pursuit of field confinement to improve interference effects), it is desired to narrow its corresponding hysteresis cycle, which results in energy losses associated to magnetizing and demagnetizing the core.

With the evolution of MEMS processes and techniques it is feasible to develop fully integrated inductors on silicon using ferromagnetic materials to increase their inductance, and thick conductors to reduce the resulting ESR. In this area, interesting works from C.H. Ahn can be consulted (good examples appear in [39] and [40]), and many other authors have developed integrated inductors by means of non-standard CMOS processes [41–49], that we include in the same group as designs based on MEMS. Recent works from Zou [50] and Musunuri [51] make use of MEMS techniques to develope vertical spiral inductor. Some work has been done in the development of off-chip microinductors, specially designed for low power converters application where the rest of the converter components are inside the chip, that should be mentioned here, too [52, 53]. Although all these kinds of integrated inductors may fulfill the aforementioned requirements, their main disadvantage is that, regarding their non-standard CMOS implementation, it is very difficult (and expensive) to include them in the mature CMOS processes, which is the purpose of the present work in order to develop PSOCs. Additionally, in some cases the used ferromagnetic materials loose their higher relative permeability as switching frequency increases [54], which is a situation to be considered in this work.

On the other hand, successful research has been carried out in the integration of inductors in standard CMOS process towards RF circuits applications. In this field a clear subdivision appears when developing on-chip inductors: those based on the use the metal layers provided by the standard CMOS process, and interesting research on the inductive effects of the bonding wires.

In the first case, inductors usually make use of the top metal layers, since commonly they provide the thickest metal paths (which reduces the ESR) and, being the most separated from the substrate, their coupling capacitances are reduced (which improves their high frequency performance). Many works on this approach can be found in the literature both about modeling them or on how to optimize their design [37, 38, 55–60].

In case of bonding wire based inductors, in spite of not explicitly belonging to the CMOS fabrication process, they are present in any integrated circuit, since they allow signals to communicate in and out of the chip. Hence, they can be considered as part of the standard CMOS process. Referred to this kind of inductors, some works exist about their modelling: [61–65]. Indeed, in many cases they are modeled because the pursuit is precisely to get rid off their inductive behavior because it affects signal integrity. Nevertheless, in some works they are intended to be used as integrated inductors [66–69]. In fact, Mertens and Steyaert used bonding wire inductors to develop an integrated Class-E power amplifier for RF applications [70].

Despite the consolidated success of inductor integration for RF applications (due to the requirements of monolithic RF transceivers, nowadays deeply present on the market), their requirements do not match those needed for the integration of switching power converters, previously listed. To sum up, they provide low parasitic stray capacitances [71] and good inductive behavior in the frequency range above 1 GHz, but usually present low inductance value, and too high ESR.

3.1.2 Proposed Integrated Inductor Implementation

After the revision of the state of the art from the aforementioned references, it is proposed to implement an inductive structure based on the bonding wires, which are always present in any chip, that provides a higher inductance that the one provided by designs for RF applications. In particular, a spiral with bonding wires as paths rather than top-metal layer paths is proposed. Obviously, since only straight segments of bonding wire are possible, any corner of the chosen geometric structure should be implemented by means of standard bonding pads.

Several advantages are expected with this implementation approach:

- Since the usual bonding wire is thicker than the top-metal of standard CMOS processes, a reduced ESR (R_L) is expected.
- Low parasitic coupling capacitance to substrate, because of the large distance of bonding wires above the silicon substrate. Just the bond pads are relatively close to the substrate, resulting in the main source of coupling capacitance to substrate.
- It is fully compatible with standard CMOS processes.
- Due to technological constraints that preclude the placement of two adjacent bonding wires very close each other, the expected intercoils coupling capacitance should be relatively small.
- Indeed, its area occupation may be highly reused to place other converter components (such as the output capacitor and power transistors) underneath the inductor, because only the pads really occupy area unuseful for other components.
- The lack of ferromagnetic core results in no switching losses in the inductor.

Unfortunately, this inductive structure also presents some disadvantages to be taken into account:

- Usually, the distance between two adjacent bonding wires is relatively large due to fabrication limitations and to prevent eventual short circuits. This results in a lower inductive density because the inner spiral turns become much smaller than the outer ones. This point will be clarified afterwards by analytical expressions.
- Despite no additional technological step is required in the fabrication process, the uncommon bonding wire structure inside the chip still requires some additional programming of the bonding wire machine.

Having stablished how to implement the inductive structure and its main characteristics, the next design step is to select a geometric shape for the spiral to be implemented in order to potentiate the desired specifications (i.e. high inductance density and low ESR).

Consequently, an equilateral triangular shape is proposed because this is the geometric shape that offers the lowest number of corners, which increase the total ESR since the bonding pads (used to implement the corners between two consecutive bonding wires) are implemented by means of the process metal layer (that exhibits higher resistivity than bonding wires). In Fig. 3.1 the main parameters used to describe a single triangular coil are observed, where $s_{ext} = 2a$ is the tringle side length (afterwards, it will become one of the main inductor design variables) and r is the radius of the bonding wire used to build up the coil.

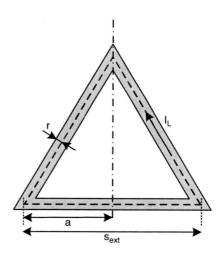

Fig. 3.1 Single equilateral triangular coil parameters

The corresponding inductance value as a function of those geometric parameters is provided by (3.1) from the Biot-Savart relationship between the magnetic field (**B**) and the current (I_L) flowing through the wire (the complete detailed analysis is discussed in the Appendix A).

$$L = \frac{\sqrt{3}\mu}{\pi}\left[\frac{r - \sqrt{3}a}{2}\ln\left(\frac{2k_{ar} + \sqrt{3}a - 3r}{2k_{ar} - \sqrt{3}a + 3r}\right) + \right.$$

$$+ \left(r - \sqrt{3}a\right)\left[\text{arctanh}\left(\frac{\sqrt{3}a}{2k_{ar}}\right) + \text{arctanh}\left(\frac{3r - 2\sqrt{3}a}{2k_{ar}}\right)\right] - k_{ar} + \sqrt{3}r -$$

$$- \frac{r}{2}\ln\left(\frac{2k_{ar} + 2\sqrt{3}a - 3r}{(3 + 2\sqrt{3})r}\right) - r\left[\text{arctanh}\left(\frac{\sqrt{3}}{2}\right) - \text{arctanh}\left(\frac{\sqrt{3}a}{2k_{ar}}\right)\right]\right]$$

$$\tag{3.1}$$

$$k_{ar} = \sqrt{3a^2 + 3r^2 - 3\sqrt{3}ar} \tag{3.2}$$

The results from (3.1) are shown in Fig. 3.2, where the obtained inductance is depicted as a function of s_{ext}. These results are also contrasted with those produced by the simplified heuristic expression from Grover inductance calculations [72].

$$L = 6 \times 10^7 s_{ext}\left[\ln\left(\frac{s_{ext}}{r}\right) - 1.40546 + \frac{\mu_r}{4}\right] \tag{3.3}$$

where μ_r is the relative permeability of the wire material, which is 1 in most cases.

As observed in the figure, there is a good matching between the results from both expressions, although Grover results are slightly more optimistic. However, the simplicity of Grover's expression makes it very handy in first approaches.

Subsequently, it is required to contrast both the inductive density and the ESR resulting from the triangular shape, with the same characteristics from several different geometric shapes. Consequently, inductance as a function of occupied area for several regular geometric shapes (triangle, square, pentagon, hexagon and octagon)

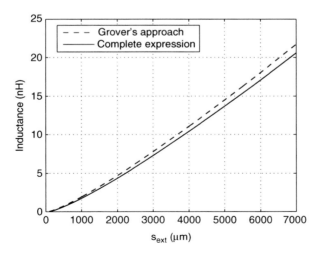

Fig. 3.2 Single equilateral triangular inductance. Comparison with Grover's expression results

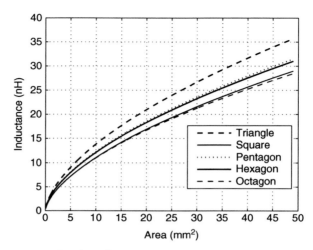

Fig. 3.3 Inductance as a function of occupied area comparison for different regular geometric shapes (according to Grover's expressions)

is compared in Fig. 3.3 (in this case, inductance expressions for all the shapes where extracted from [72]).

As a consequence, it is clarified that the triangular shape offers the highest inductive density (L/A_L) among all the considered shapes (which due to technological limitations, they all must be polygonal shapes).

Furthermore, it is very interesting to evaluate which is the geometric shape that produces the lowest ESR, for the same inductance value (i.e. the lowest R_L/L ratio). Hence, R_L is evaluated for any different shape, after the obtention of the s_{ext} that provides the desired inductance value.

$$R_L = n_{sides} \left(\frac{s_{ext}}{\sigma \pi r^2} + R_{bp} \right) \tag{3.4}$$

In (3.4), n_{sides} is the number of sides of any considered shape, $\sigma = 41 \times 10^6$ $Siemens/$m, the bonding wire radius $r = 12.5\,\mu$m and the bonding pad resistance $R_{bp} = 13\,$mΩ.

With this information (extracted from a standard CMOS process), the produced ESR for different inductance values is contrasted among the considered geometric shapes in Fig. 3.4. In this figure, it becomes clear that the lowest ESR is obtained by implementing a triangular inductor.

To conclude, it is found that the equilateral triangular shape is the most convenient to implement a planar spiral inductor for switching power converter integration purposes, since it best fulfils the main requirements for this application.

Still, according to results from Fig. 3.3 for triangular shape, its inductive density is rather low. Thus, an spiral with several number of turns could be required to obtain the desired inductor, in order to increase its effective inductive density.

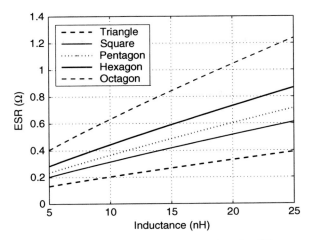

Fig. 3.4 ESR as a function of the desired inductance comparison for different regular geometric shapes

As a consequence, the inductance value of a triangular spiral should be evaluated. In this case, not only self-inductance for any coil is necessary, but also mutual inductance between all the coils must be computed. This can be expressed as follows:

$$L = \sum_{j=1}^{n_L} \sum_{i=1}^{n_L} L_{ij} \qquad (3.5)$$

Previously, the expression for all the terms where $i = j$ (self-inductance) was found in (3.1). Moreover, it is observed that referring to the terms related to the mutual inductance ($L_{ij} \rightarrow i \neq j$), $L_{ij} = L_{ji}$, because of reciprocity. Thus, by means of iterations the total inductance of the spiral can be found (for a more complete explanation and details, see Appendix A). The results for the total inductance of a triangular spiral are shown in Fig. 3.5. This 3D figure shows the obtained inductance as a function of the outer triangle side length, and the number of turns. It is noted that not all the points arising from the axis projection are depicted. This is because in some cases the number of turns is not feasible inside the triangle, given a positive distance between two adjacent turns (whose minimum value is specified by the bonding wire technology). In fact, the distance between turns (denoted p) has a notable impact on the resulting inductance. To illustrate this effect with an example, Fig. 3.6 depicts the inductance of a given spiral ($s_{ext} = 3.5$ mm, $n_L = 3$ and $r = 12.5 \,\mu$m), as a function of the distance between turns p.

As observed in Fig. 3.6, it is convenient to place the bonding wire segments as close as possible from each other to increase the resulting inductance with the same s_{ext} value (i.e. to increase its inductive density).

In order to clarify the inductive density evolution of the proposed inductors, the occupied area is evaluated in Fig. 3.7, from (3.6). As expected, occupied area is only

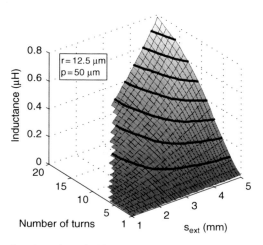

Fig. 3.5 Inductance of a triangular spiral inductor as a function of the number of turns and the outer side length

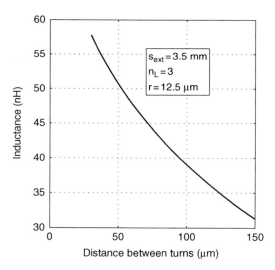

Fig. 3.6 Inductance of a triangular spiral inductor as a function of the distance between turns ($s_{ext} = 3.5\,\text{mm}$, $n_L = 3$ and $r = 12.5\,\mu\text{m}$)

dependent on the s_{ext} value, producing higher L/A_L ratio as the number of turns is increased.

$$A_L = \frac{\sqrt{3}s_{ext}^2}{4} \qquad (3.6)$$

Following the inductance and area evaluation of the proposed triangular spiral, it is necessary to calculate the ESR of the overall spiral inductor. Both number of

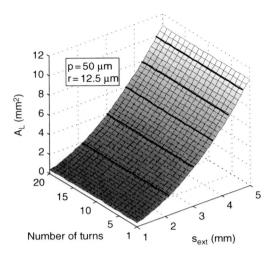

Fig. 3.7 Occupied area by the triangular spiral inductor as a function of the number of turns and the outer side length

corners as well as the overall bonding wire length contribute to the ESR value (R_L), as stated in (3.7).

$$R_L = R_{bp} (3n_L + 1) + 3\zeta n_L (s_{ext} + q (1 - n_L)) \tag{3.7}$$

$$q = \sqrt{3}p \tag{3.8}$$

where $\zeta = \frac{1}{\sigma \pi r^2}$ is the wire resistivity and q is half of the side reduction for any inner turn added to the spiral, provided that s_{ext} remains constant (for further details on the inductor overall ESR computation, see Appendix A). As for inductance calculations, the total inductor ESR as a function of the number of turns and the outer side length is shown in Fig. 3.8.

Results from Fig. 3.8 reveal high ESR values for power converter integration purposes, despite the use of bonding wires and triangular shape selection. However, it should be noted that inductance results show relatively high inductance values (relative to other integrated inductors).

With respect to a suitable inductor design, it is interestingly noted that several pairs (n_L, s_{ext}) provide the same inductance value (given by the contour lines of Fig. 3.5), even though they do not result in the same amount of occupied area and R_L value. Inevitably, this implies that among the several designs that produce the same L value some will perform better in the considered framework. As a consequence, an optimization procedure should be carried out regarding the inductor design.

In this context, the work developed in [38] should be consulted, where an analytical method (based on geometric programming) was presented to optimize inductors designs (towards RF applications, in that case), and interesting results were obtained. However, it required to find the coefficients of some model expressions by

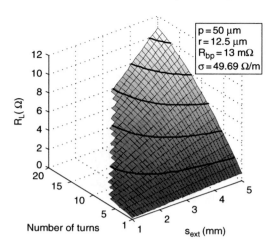

Fig. 3.8 ESR of a triangular spiral inductor as a function of the number of turns and the outer side length

experimentally characterizing a batch of test structures, which was not possible to carry out in our work.

Essentially, it is desired an inductor design which provides the desired inductance while reduces the occupied area as well as the ESR. To fulfil these requirements, a merit figure particular for the inductor design is defined in (3.9).

$$\Gamma_L = \frac{1}{[A_L]^{\gamma_{LA}} [R_L]^{\gamma_{LR}}} =$$
$$= \left[\left[\frac{\sqrt{3}s_{ext}^2}{4} \right]^{\gamma_{LA}} \left[R_{bp}\,(3n_L + 1) + 3\zeta n_L \left(s_{ext} + \sqrt{3}p\,(1 - n_L) \right) \right]^{\gamma_{LR}} \right]^{-1}$$

(3.9)

The exponents affecting both terms of (3.9) can be used to modify the optimization results by giving more strength to area or resistance minimization (in this work: $\gamma_{LA} = 1$ and $\gamma_{LR} = 10$).

Since Γ_L is defined as the inverse of the product of two undesired characteristics, the optimization procedure should maximize it.

The first step is to select all the (n_L, s_{ext}) pairs that provide the desired inductance value. Being the spiral inductance evaluation an iterative process, there is not an analytical expression that provides the value of the desired inductance, and hence numerical methods must be applied. Furthermore, although s_{ext} is a continuous magnitude, in this work n_L is defined to exist just in natural values. As a result, it might be that none of the considered pairs of values produce the exact value of the desired inductance. Consequently, a narrow tolerance range around the desired value is defined and any design inside it is considered to be valid.

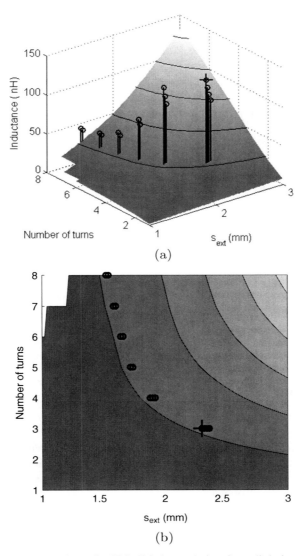

Fig. 3.9 Optimization procedure of a 27.3 nH inductor design. In **a** all designs that provide an inductance of 27.3 nH (±3%) are marked with vertical lines whose height is a qualitative representation of its corresponding merit figure value. **b** is a top view of the same results. In both cases, the optimized design is marked with a *cross*. The values of the used parameters can be found in Table 3.1

To carry on, (3.9) is evaluated for any of the selected (n_L, s_{ext}) pairs and the one that provides the maximum merit figure value is considered as the optimized inductor design.

In Fig. 3.9, an example for a design providing a 27.3 nH inductance can be observed. In the 3D view, selected designs are marked with superimposed vertical lines, whose height is a qualitative representation of the corresponding merit figure value. Additionally, contour lines are depicted for each 25 nH. Figure 3.9b is just a top representation, where the position of the selected designs is better identified. In both pictures, the optimized design is marked with a cross (corresponding to the maximum merit figure value, as observed in Fig. 3.9b).

Table 3.1 summarizes the main characteristics of the optimized design (as well as the parameters values), where it can be observed the effect of the tolerance around the desired inductance value, as a 2.2% error was obtained in the final obtained inductance value.

Table 3.1 Main characteristics and parameters of the optimized inductor design

Technical parameter	Value
Desired inductance	27.3 nH
Tolerance on desired value	3%
Bonding pad resistance R_{bp}	13 mΩ
Bonding wire resistivity ζ	49.69 Ω/m
Bonding wire radius R	12.5 μm
Distance between bonding wires p	50 μm
Area coefficient γ_{LA}	1
Resistance coefficient γ_{LR}	10
Characteristic	Value
Final inductance L	26.7 nH
Outer side length s_{ext}	2.3 mm
Number of turns n_L	3
Occupied area A_L	2.3 mm^2
ESR R_L	1.08 Ω

In case of bonding wire structures, it should be noted that neither the path width (bonding wire diameter is fixed by the technology), nor the separation between turns (kept at minimum possible, according to results from Fig. 3.6) are design variables, while they are common in metal-layer planar spiral designs.

In summary, the proposed design procedure of an inductor to be integrated is treated as a *design space exploration*, such as the exposed in Chap. 2, but particularized for the inductor design. In the same way, the required characteristics of the inductor (ESR and occupied area, (3.7) and (3.6)) are considered as *performance factors* and accordingly modeled to include them in a merit figure definition to be maximized (in the inductor design case, (3.9)), and then the design space exploration is carried out. The result of this process is a complete set of design parameter and characteristics (Table 3.1).

Having concluded the modelling and design procedure for the inductor, a finite-element model (Fig. 3.10) and simulation of the proposed structure was carried out to validate the proposed expressions. In the developed model, expected materials to be found in an integrated environment are included, such as silicon substrate, field oxid and the aluminum bonding pads.

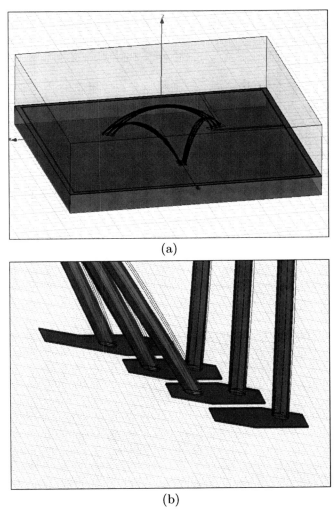

(a)

(b)

Fig. 3.10 Inductor model developed for finite element simulations. In **a**, all the considered materials are included apart from the spiral itself. In **b**, the detail of one of the corners with the required bonding pads is presented

Magnetostatic simulation results of the finite element model yield a $L = 34\,\mathrm{nH}$, $R_L = 1.08\,\Omega$ inductor. Additionally, the resulting inductance as a function of the frequency is shown in Fig. 3.11, where it is found a relative good matching between the expected inductance and the results for a frequency range below 200 MHz, which is reasonable for the considered applications. Simulation for low frequency was not possible, since the used simulator (*HFSS*) is intended to simulate RF designs.

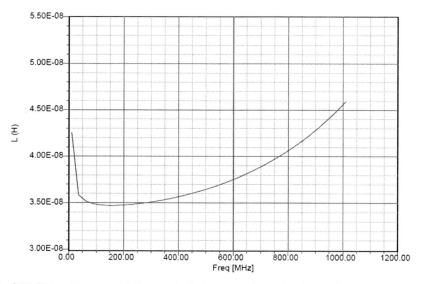

Fig. 3.11 Finite element simulation results. Inductance value as function of frequency

Fig. 3.12 Finite element simulation results. Spatial magnetic field distribution on silicon surface

Finally, magnetic field density on the silicon surface for an inductor current of $i_L = 200\,\text{mA}$ is depicted in Fig. 3.12. It is found that major concentrations of magnetic field on the silicon occur around the corners. Thus, sensible circuit components should not be placed near the corners to avoid high interferences.

However, further in-depth analysis and simulations would be necessary to study these effects.

3.2 MOS Capacitor Model

3.2.1 State of the Art

Typically, on-chip capacitors have been developed for analog signal processing applications (e.g. switched-capacitor filters or sample-and-hold circuits, [73]). Such applications require on-chip capacitors that present good matching (i.e. capacitors with similar characteristics) and linearity (i.e. their capacitance do not change when the applied voltage changes). However, due to low current management, they do not require very low series resistance, neither high capacitive density, provided that no high capacitance values are needed.

RF applications also make use of integrated capacitors and, to the aforementioned requirements, add the need for high values of Q factor and high self-resonance frequency (both indexes describe the quality of its capacitive behavior at high-frequency), since they should operate in the high-frequency band.

On the other hand, on-chip power management applications such as on-chip power supply decoupling and on-chip switching power converters present different main requirements: high capacitive density is fundamental (since high capacitance values are in principle needed), and very low equivalent series resistance (from now on, ESR), in order to reduce energy losses and their equivalent impedance (which in case of power converters results in higher output ripple value). Nevertheless, such kind of applications neither require very good matching (usually no capacitors are needed to be paired), nor precision on its value.

Because of the planar environment, that constraints the design possibilities, the most common implementation of on-chip capacitors relies on the horizontal parallel plates conception (in which energy is stored in the vertical electric field). This is really straightforward since the resulting capacitance depends on the plates area, their separation and the permeability of the material between them (commonly, it is silicon dioxide). Usually, this yields high linearity capacitors, although the capacitive density is rather low (because of the relatively high separation between plates, provided that no special material is inserted between them). Their ESR depends directly on the plates material: it used to be polysilicon (resulting in high ESR values), but most recent technologies provide metal plates (mainly aluminum) capacitors. Another interesting feature that stems from parallel plate capacitors is that both plates are intrinsically floating, thus they do not need to be connected to any particular voltage level.

In this approach, a very interesting proposal from Samavati to use fractal capacitors in order to increase the capacitive density (by taking benefit from the lateral electric field) can be consulted [74]. More recently, a more general comparison between capacitive structures that take benefit from the lateral electric field (including fractal capacitors), and parallel plate capacitors was published by Aparicio [75].

In some cases, parallel plate capacitors have been used to power management applications, too. In [76], *poly-poly* capacitors are used in an on-chip charge-pump implementation, where floating plates are needed. Nevertheless, this is not a representative case for the purpose of this work, since charge-pumps usually do not require such large capacitors.

Other kinds of on-chip capacitors used in RF and analog processing applications are based on the parasitic capacitance of some of electronic devices. Some are based on the parasitic capacitor of a reverse biased PN junction, just because of its non-linearity. This particularity leads to electrically variable capacitors, that become useful when tunning RF resonators or filters [77]. Also the non-linear behavior of the parasitic gate capacitance of MOS transistors has been proposed to be used as a varactor using the body voltage to tune its value [78]. Both kind of on-chip capacitors present the disadvantages of high series resistance, and the need for a bias voltage to generate the capacitive behavior.

The design of on-chip power supply decoupling capacitors (which roughly demand similar characteristics than in case of switching power converters application) has been widely addressed by many authors. In [79–81] the need for on-chip decoupling capacitors and their characteristics are covered as a part of the global supply system, and interesting calculations and placement considerations are exposed. However, no specific technical solutions are provided.

In [82, 83], Schaper et al. proposed a method for implementing on-chip capacitors for supply decoupling purposes that yields very low equivalent series inductance and resistance. Unfortunately, it was not fully standard CMOS compatible, and presented low capacitive density.

In this work, the output capacitor of integrated switching power converters is proposed to be implemented by means of the MOSFET gate capacitor (i.e. the gate terminal becomes the top capacitor plate, whereas the bottom plate is formed by the transistor channel when drain and source are shorted). Specifically, NMOS transistors are used because of the higher majority carriers mobility. In the following, main considerations about this election are listed (including advantages and disadvantages):

- The most important feature of this kind of capacitor is that it presents the highest capacitive density because the gate silicon dioxide insulator is the thinnest layer between two conductors of an standard CMOS IC.
- The dependence of the capacitor value on the applied voltage is not an important drawback, as long as the applied gate voltage is high enough to generate the transistor channel (Fig. 3.13). Hence, if a wide span of the converter output voltage is desired, this issue should be considered. Furthermore, in case of highly non-linear behavior, it should be studied its effect on the output voltage ripple.
- Since NMOS transistors are used as MOS capacitors, drain and source should be connected to the same voltage than the surrounding well or substrate (which usually is GND), to short-circuit the parasitic capacitor and avoid forward-biasing the corresponding PN junctions.
- Capacitor ESR is expected to be exceedingly high (as regards power efficiency and output ripple), because of the polysilicon and channel resistivities. Additionally, channel resistivity depends on the applied gate-to-source voltage. Inevitably, specific and accurate design strategies are required to reduce the ESR.

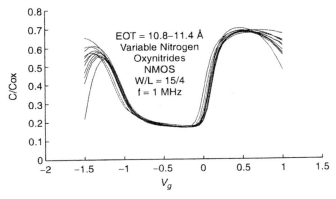

Fig. 3.13 Qualitative representation of the parasitic gate capacitor variation as a function of the gate voltage (reprinted from [5]). Once the V_T threshold is exceeded (and the channel is created), the capacitance value remains rather constant

- In case of technological nodes beyond 0.25 μm or 0.18 μm (and the consequent reduction of the gate insulator thickness), leakage current through the gate insulator due to tunnel effect could result in energy losses no longer negligible [5].
- If possible, *Low-V_t* transistors are preferred to implement the MOS capacitor because the required gate voltage for channel formation is lower, which, additionally, yields a lower channel resistance for the same gate voltage value.

3.2.2 Proposed MOS Capacitor Design Procedure

As mentioned in the previous section, one of the main disadvantages of the MOS capacitors (MOSCAP, from now on) is their relatively high ESR, which not only increases power losses, but may result in an important increase of the output voltage ripple. Therefore, it becomes fundamental to model the corresponding ESR to predict its value and, necessarily, reduce it by means of a properly optimized design.

3.2.2.1 ESR of a Single MOS Capacitor

The first issue to consider when modelling the MOSCAP ESR is the distributed nature of the gate capacitance and the parasitic resistances from gate polysilicon, as well as the transistor channel. This is a matter that has been addressed by different authors in the bibliography. Especially, works from Larsson [6] and Jin [7], are very useful when attempting to model the MOSCAP ESR (although they were developed for RF application purposes).

Larsson's Model

This model only considers the distributed channel resistance as shown in Fig. 3.14, whose equivalent circuit is depicted in Fig. 3.15a [6]. In this scheme, c_n is the

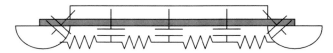

Fig. 3.14 Distributed channel resistance of a MOSCAP [6]

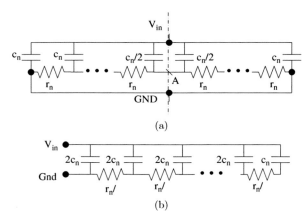

Fig. 3.15 Equivalent circuit of the channel resistance and capacitance of a MOSCAP. **a** Direct physical to circuit translation. **b** Simplified version after folding through the central point A [6]

capacitance per unit length, whereas r_n is the resistance per unit length. Since due to circuit symmetry it can be assumed that no current flows through node A, the circuit can be folded, resulting in the open-ended transmission-line model in Fig. 3.15b, whose impedance is given by:

$$Z_{dist} = \sqrt{\frac{\frac{R}{4}}{sC}} \coth\left(\sqrt{\frac{sRC}{4}}\right) \tag{3.10}$$

where R is total channel resistance, and C is total gate capacitance ($C = C_{ox}WL$, being W and L the gate length and width, respectively).

The series expansion of (3.10) results in:

$$Z_{dist} = \frac{R}{4x}\left(\frac{1}{x} + \frac{x}{3} - \frac{x^3}{45} + \ldots\ldots\right) \approx \frac{1}{sC} + \frac{R}{12} - \frac{sCR}{720}R \tag{3.11}$$

where $x = \sqrt{sRC/4}$. By comparing to the impedance of a lumped series RC circuit (3.12), it can be noted that whereas the expression of the capacitance is coincident, the resistance of the distributed model is frequency-dependent and corresponds to the $1/12$ of the lumped model.

$$Z_{lump} = \frac{1}{sC} + R \tag{3.12}$$

Hence, the ESR of a MOS capacitor is (channel resistance only):

$$R_{dist} = \frac{R}{12} \qquad (3.13)$$

$$R = \left. \frac{dV_{ds}}{dI_{ds}} \right|_{V_{ds}=0} \qquad (3.14)$$

Jin's Model

More recently, Jin et al. [7] presented a more detailed model that includes the gate polysilicon resistance as well (shown in Fig. 3.16), which is given by:

$$R_G = R_{poly}\left(\alpha\frac{W}{L} + \beta\right) \qquad (3.15)$$

where R_{poly} is the polysilicon square-resistance, β is the external resistance (due to connection path) and α is either $1/3$ or $1/12$, if the gate is connected from one or both sides (short-circuiting both extremes), respectively.

Fig. 3.16 Distributed channel and gate resistances of a MOSCAP [7]

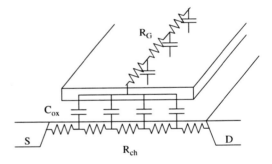

Apart from the static resistance R_{st} (3.16) (equivalent to the one modeled by Larsson), this model accounts for an additional source of channel resistance due to changes in channel charge R_{ed} (3.17), hence being the complete channel resistance R_{ch} composed of two parallel terms (3.18).

$$R_{st} = \left. \frac{dV_{ds}}{dI_{ds}} \right|_{V_{ds}=0} = \frac{L}{\mu_N C_{ox} W (V_{gs} - V_{TN})} \qquad (3.16)$$

$$R_{ed} = \frac{L}{\eta W \mu_N C_{ox} V_{term}} \qquad (3.17)$$

$$R_{ch} = \frac{R_{st} R_{ed}}{\gamma (R_{st} + R_{ed})} = \frac{L}{\gamma W \mu_N C_{ox} (\eta V_{term} + V_{gs} - V_{TN})} \qquad (3.18)$$

where η is a technology dependent constant (and usually is 1), $\gamma = 12$ provided that the transistor operates in active region (which is the case for $V_{ds} = 0$, i.e. both extremes of the transistor channel are short-circuited), and hence matches with (3.10).

The complete model for the ESR of a MOS capacitor resorts to the previous models, which, when neglecting external resistance (it is assumed a perfect connection in both terminals) and frequency dependence, is given by:

$$ESR = R_{ch} + R_G = \frac{1}{\gamma \mu_N C_{ox}(\eta V_{term} + V_{gs} - V_{TN})} \frac{L}{W} + R_{poly}\alpha \frac{W}{L} \qquad (3.19)$$

Due to the complementary dependence on MOSFET transistor aspect ratio (W/L) of the two terms in expression (3.19), the ESR will exhibit an absolute minimum for a certain aspect ratio:

$$\frac{dESR}{d(W/L)} = 0 \longrightarrow \left. \frac{W}{L} \right|_{opt} = \sqrt{\frac{1}{\gamma \mu_N C_{ox}(\eta V_{term} + V_{gs} - V_{TN})R_{poly}\alpha}} \qquad (3.20)$$

And this optimum aspect ratio produces the minimum attainable capacitor ESR for a given set of technological parameters:

$$ESR_{min} = 2\sqrt{\frac{R_{poly}\alpha}{\gamma \mu_N C_{ox}(\eta V_{term} + V_{gs} - V_{TN})}} \qquad (3.21)$$

This value is obtained for the case of equal gate and channel resistances ($R_G = R_{ch}$). Note that the minimum attainable value of ESR is dependent on both technological parameters, and the voltage applied to the MOSCAP. Furthermore, it is independent of the capacitance value since this value relies on the area of the MOSCAP gate, but not on its aspect ratio.

Figure 3.17 shows the evolution of the capacitor ESR as a function of its aspect ratio, for example process parameters (corresponding to a 0.25 μm process). From these results, it is noted that a square-shaped MOSCAP would result in a extremely high ESR, due to the high channel resistance value. Thus, it becomes of capital importance to implement MOSCAPs with an optimized aspect ratio.

3.2.2.2 MOSCAP Matrix Structure

The results from Fig. 3.17 expose that even when the MOSCAP presents an optimum aspect ratio to reduce its ESR, this results exceedingly high to satisfy specifications for most of the power applications.

It is well known that a usual strategy to reduce ESR using discrete components considers parallel connection of several capacitors. Caution should be taken when extending this concept to the VLSI technology, since the direct snake-shaped parallel connection of several optimum-sized transistors (Fig. 3.18a) results in the parallel

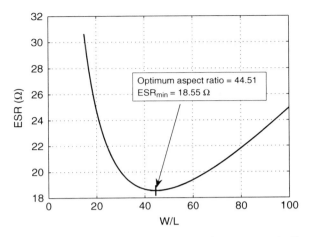

Fig. 3.17 MOSCAP ESR evolution as a function of the transistor aspect ratio. Parameters values: $V_{term} = 25\,\text{mV}$, $C_{ox} = 6.27\,\text{fF}/\mu\text{m}^2$, $\mu_N = 376\,\text{cm}^2/(\text{sV})$, $V_{TN} = 0.17\,\text{V}$ (Low-V_T transistor are considered), $R_{poly} = 2.5\,\Omega$, $\gamma = 12$, $\alpha = 1/12$

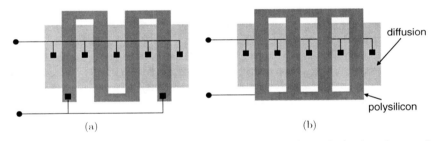

Fig. 3.18 Connection of several MOSCAPs to reduce the total ESR. **a** *Snake-shaped* connection results in an equivalent non-optimum design. **b** Parallel connection of gates, drains and sources, results in true parallel connections of channel resistances and gate resistances

connection of corresponding channels whereas the associated polysilicon gates are connected in series, yielding a non-minimum ESR (equivalently, as becomes clear in Fig. 3.18a: the transistor increases its aspect ratio and loses optimality as regards ESR). In fact, this kind of connection results in the parallel connection of the channel resistances, whereas gate resistances become series connected, thereby producing high ESR value.

The connection scheme shown in Fig. 3.18b, produces the parallel connection of both gates and channels terminals, respectively. As a result, the achieved ESR is (provided that transistors are of optimum aspect ratio):

$$ESR = \frac{ESR_{min}}{n} \tag{3.22}$$

where n is the number of MOSCAP in parallel connection.

In the following, a matrix structure that connects multiple MOSCAPs by the use of the gate polysilicon layer plus the 3 bottom level metal layers of the considered process is presented, and the corresponding area considerations are provided. The capacitance of a MOS capacitor depends on the transistor gate area and its capacitance per unit area (C_{ox}). As shown in (3.22), the ESR depends on both technology (ESR_{min}) and the number of parallel transistors (n). Therefore, the total area of the capacitor depends on the required values of C and ESR, because a portion of the total area will be used to interconnect the transistors (which could be significant, since many parallel-connected MOSCAPs could be required to achieve an ESR value low enough). Following this reasoning, two types of areas are distinguished: the *capacitive area* (total gate area, A_{Cc}) and the *routing area* (area spent to interconnect the transistors, A_{Cr}). Capacitive area is proportional to the number of transistors and their dimensions (W and L), and directly related to the required capacitance value:

$$A_{Cc} = \frac{C}{C_{ox}} = nWL \tag{3.23}$$

Routing area will depend on the number of transistors, their dimensions and the overhead area per transistor used to interconnect them. Since the overhead area depends exactly on the particular structure used to arrange the transistors and connect them, in the following the herein proposed matrix structure is presented.

Transistors are arranged in a matrix modular structure (Fig. 3.19). Gates are interconnected by means of a polysilicon sheet where only contacts *metal1-difussion* open the necessary holes (Fig. 3.19a). In parallel with each row of transistors, there is a *metal1* strip (and their corresponding contacts to polysilicon) which reduces the resistance between adjacent gates (Fig. 3.19a). Afterwards, all these *metal1* strips are interconnected using a wide *metal2* sheet (Fig. 3.19c), which becomes the capacitors gates terminal. As regards drains and sources, they are connected using strips of *metal1* over transistors (with W width), which are, in turn, interconnected at the ends of each row, by means of *metal1* and *metal2* strips (and the corresponding vias *metal1-metal2*), as exposed in Fig. 3.19b. Finally, the two *metal2* strips are short-circuited using a sheet of *metal3*, which is the capacitor terminal corresponding to the drains and sources (Fig. 3.19d).

In order to properly create the transistors channels, higher voltage should be connected to gates terminal (since NMOS transistors are used to implement the MOSCAP structure). Usually, drains and sources terminal should be grounded to short-circuit the parasitic substrate coupling capacitor.

Analyzing an unitary cell of the proposed structure (detailed in Fig. 3.20), the overhead area per transistor can be determined as a function of gate dimensions, source and drain contacts dimensions, and the inter-element distances according to layout design rules (determined by the technological process). Then, an expression for the routing area of the overall structure is obtained:

$$A_{Cr} = n(a(L+b) + Wb) \tag{3.24}$$

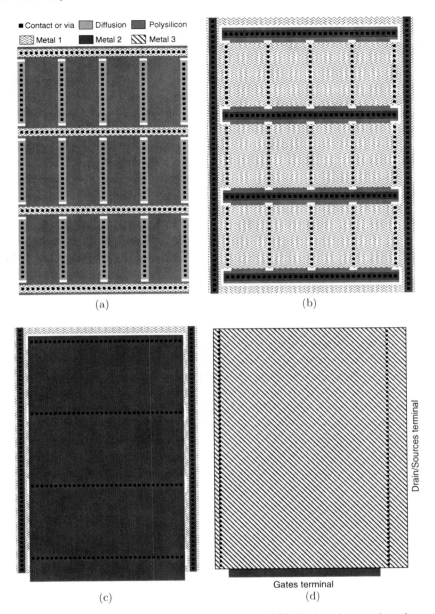

Fig. 3.19 Proposed matrix structure to connect several MOSCAPs in order to reduce the total capacitor ESR

Fig. 3.20 Single cell
dimensions of the MOSCAP
structure

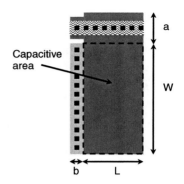

where b and a represent length and width overhead, respectively. This expression
does not take into account the area spent in peripheral interconnections, but it can
be assumed that it is not significant in front of the area spent to interconnect a large
number of cells. According to layout design rules for the considered 0.25 μm technology, $a = 0.8\,\mu\text{m}$ and $b = 0.8\,\mu\text{m}$.

A benefit from the modular structure of MOSCAP matrix herein presented is that
it is more suitable to fill the interstices between the circuits that appear in many chip
layouts. Therefore, it can be used to take profit from those spaces to distribute the
output capacitor of an integrated switching converter (as proposed by Abedinpour
in [84]), or even for power supply decoupling.

In the following, a resistive model of the proposed structure is provided in order
to take it into account when determining the total ESR of the overall MOSCAP. To
develop the model, the whole structure is divided in two different circuits: the gate
circuit and the channel circuit; and both of them are separately analyzed.

Before that, note that throughout the analysis it will be supposed a matrix structure with A columns and B rows (Fig. 3.21).

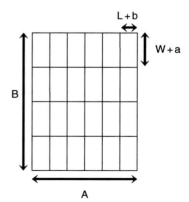

Fig. 3.21 Columns and rows
subdivision of the MOSCAP
matrix structure

The Gate Circuit

To analyze the gate circuit, the ESR of a single MOS capacitor is split in two parts: one due to the gate polysilicon (R_G) and the second one due to the channel (R_{ch}). Only R_G will be attached to the gate circuit. In Fig. 3.22 a transversal section of the transistor with the interconnections corresponding to the gate circuit, is shown.

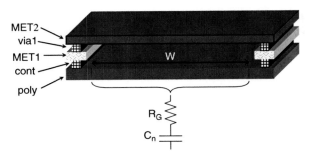

Fig. 3.22 Physical structure of the MOSCAP matrix gate circuit

As it can be seen, the gate polysilicon (the resistance of which has been previously determined) is short-circuited by means of the *metal2* layer at the transistor boundaries. In this case the interconnection parasitic resistance can be divided into the one that corresponds to the *metal2* layer (r_{m2}), and the two connections between this layer and the polysilicon (r_{pm2}), as stated in Fig. 3.23.

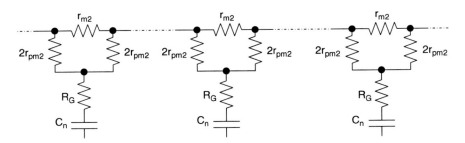

Fig. 3.23 MOSCAP gate circuit electrical scheme

In Fig. 3.23 the interconnection parasitic resistances between *metal2* and polysilicon layers have double their actual value because two adjacent cells share the same interconnection (thus, their parallel connection results in r_{pm2}).

The r_{m2} value is proportional to the *metal2* square resistance (R_{MET2}) and the aspect ratio of the cell:

$$r_{m2} = R_{MET2} \frac{W}{L+b} \tag{3.25}$$

On the other hand, r_{pm2} results from the series connection of several parallel-connected vias, and several parallel-connected *poly-metal1* contacts:

$$r_{pm2} = \frac{R_{cont}}{(L+b)\delta_{cont}} + \frac{R_{via1}}{(L+b)\delta_{via1}} \qquad (3.26)$$

where R_{cont} and R_{via1} are the resistance values of *poly-metal1* contact and *via1* (*metal1-metal2* contact), respectively. δ_{cont} and δ_{via1} are the linear densities of contacts and *via1s*.

The circuital structure in Fig. 3.23 does not allow an evaluation of the resistance that results from the connection of the multiple cells of a column. Hence, the triangle connection is translated to star connection by means of Kenelly's theorem [85]. Consequently, the connection of several cells of a column is represented in Fig. 3.24.

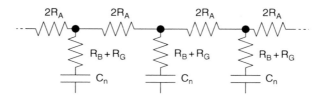

Fig. 3.24 MOSCAP gate circuit electrical scheme, after the triangle-to-star transformation

And the new resistors R_A and R_B values are:

$$R_A = \frac{2r_{pm2}r_{m2}}{4r_{pm2} + r_{m2}} \qquad (3.27)$$

$$R_B = \frac{4r_{pm2}^2}{4r_{pm2} + r_{m2}} \qquad (3.28)$$

Taking into account that the *metal2* layer along the whole column is connected from both sides, the Larsson's approach can be applied. Therefore, the circuit is folded as shown in the following procedure:

1. All the circuit branches from the scheme of Fig. 3.24, are considered as complex impedances (Fig. 3.25), with the corresponding distributed impedance expressions:

$$Z_{dist} = \sqrt{Z_1 Z_2} \coth\left(\sqrt{\frac{Z_2}{Z_1}}\right) \qquad (3.29)$$

$$Z_2 = nZ_{2n} = 2BR_A \qquad (3.30)$$

$$Z_1 = nZ_{1n} = \frac{s(R_B + R_G)C_n + 1}{sBC_n} \qquad (3.31)$$

Here, C_n is the gate capacitance of a single MOSCAP cell.

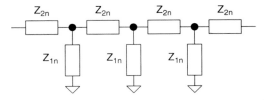

Fig. 3.25 MOSCAP gate circuit electrical scheme, generic complex impedances consideration

2. If the circuit is folded from its middle point, the Z_1 and Z_2 values are:

$$Z_2 = \frac{BR_A}{2} \tag{3.32}$$

$$Z_1 = \frac{s(R_B + R_G)C_n + 1}{sBC_n} \tag{3.33}$$

3. Expanding (3.29) by means of Taylor's series:

$$Z_{dist} = \frac{Z_2}{x}\left(\frac{1}{x} + \frac{x}{3} - \frac{x^3}{45} + \ldots\right) \tag{3.34}$$

$$x = \sqrt{\frac{Z_2}{Z_1}} \tag{3.35}$$

4. Finally, considering the three most important terms of (3.34):

$$Z_{dist} = \frac{1}{sC_nB} + \frac{R_G + R_B}{B} + \frac{BR_A}{6} - \frac{B^3 R_A^2 C_n s}{180[s(R_G + R_B)C_n + 1]} \tag{3.36}$$

Once the impedance expression of a column of cells is found, the parallel connection of A columns is considered, thus:

$$Z_{dist} = \frac{1}{sC_nBA} + \frac{R_G + R_B}{BA} + \frac{BR_A}{6A} \tag{3.37}$$

where the frequency-dependent term has been neglected. As $AB = n$, the total capacitance is represented by the first term of (3.37). Hence, the total ESR of the gate circuit is expressed as follows:

$$ESR_G = \frac{R_G + R_B}{n} + \frac{BR_A}{6A} \tag{3.38}$$

The Channel Circuit

Basically, the channel circuit analysis is carried out through a reasoning analogous to the previous one corresponding to the gate circuit. In Fig. 3.26, the longitudinal

Fig. 3.26 Physical structure
of the MOSCAP matrix gate
circuit

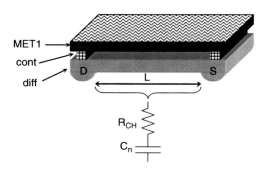

section of a single cell of the MOSCAP matrix is found, whose equivalent electrical
circuit is depicted in Fig. 3.27a, being the resistor values:

$$r_{m1} = R_{MET1}\frac{L + b}{W} \tag{3.39}$$

$$r_{dm1} = \frac{R_{cont}}{W\delta_{cont}} \tag{3.40}$$

where R_{MET1} is the *metal1* layer square resistance and R_{cont} and δ_{cont} represent the
resistance and linear density of *diffusion-metal1* contacts, respectively.

By means of the triangle-to-star connection transformation (Fig. 3.27b), the
equivalent resistor values of the star-connected circuit (which is more suitable to
further circuit calculations) become:

$$R'_A = \frac{2r_{dm1}r_{m1}}{4r_{dm1} + r_{m1}} \tag{3.41}$$

$$R'_B = \frac{4r^2_{dm1}}{4r_{dm1} + r_{m1}} \tag{3.42}$$

Fig. 3.27 MOSCAP channel
circuit scheme. **a** Direct
translation from physical
structure. **b** After the
triangle-to-star
transformation

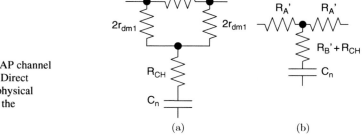

In expressions (3.41) and (3.42), the ′ symbol has been added just to distinguish these parameters from their gate circuit counterparts.

Considering both-sided connections of rows, Larsson's reasoning can be applied here too, and the provided row of cells impedance is expressed in (3.43), having neglected the frequency-dependent term.

$$Z_{dist} = \frac{1}{sC_n A} + \frac{R_{ch} + R'_B}{A} + \frac{A R'_A}{6} \qquad (3.43)$$

In contrast to the gate circuit case, the parallel connection of B rows can not be considered directly, because in this case the rows are interconnected using the *metal3* layer, which covers the whole structure, and *metal1* connections too, at the edges of each row. Hence, the corresponding circuit is the one depicted in Fig. 3.28.

In Fig. 3.28a R_x represents the *metal1-metal3* connection at the edge of each row, and R_y represents the connection between two adjacent rows through the *metal1* layer, at both edges of the rows. Figure 3.28b depicts the equivalent circuit which is obtained joining R_x parallel pairs. It is interesting to note that in case of an ideal connection on top of the $R_x/2$ resistors, no current will flow through R_y due to the equality of the voltages at their extremities; thus, their value is not necessary to be known. Considering that the external connection by means of the *metal3* layer is realized in its middle point, the R_x value provided is by (3.46).

$$r_{m3} = R_{MET3} \frac{A(L + b)}{W + a} \qquad (3.44)$$

$$r_{m13} = \frac{R_{via1}}{(W + a)\delta_{via1}} + \frac{R_{via2}}{(W + a)\delta_{via2}} \qquad (3.45)$$

$$R_x = \frac{r_{m3}}{2} + r_{m13} \qquad (3.46)$$

where R_{MET3} is the *metal3* layer square resistance, and R_{via2} and δ_{via2} are the resistance and the linear density of the *via2*'s (*metal2-metal3* contacts).

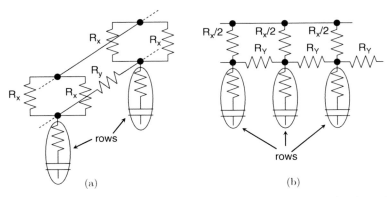

Fig. 3.28 Scheme corresponding to several rows connection. **a** Direct translation from physical structure. **b** After some simplification

According to expressions (3.44), (3.45) and (3.46), the first term of R_x depends on the matrix shape (by means of the factor A) whereas the second one is independent of it, as stated on the following expression:

$$R_x = R'_x A + R''_x = \frac{R_{MET3}(L+b)}{2(W+a)} A + \frac{1}{W+a} \left(\frac{R_{via1}}{\delta_{via1}} + \frac{R_{via2}}{\delta_{via2}} \right) \qquad (3.47)$$

Taking into account $R_x/2$ resistors, as they appear in Fig. 3.28b, now the parallel connection of B rows can be assumed. As a consequence, the total impedance of the channel circuit is:

$$Z_{dist} = \frac{1}{sC_n AB} + \frac{R_{ch} + R'_B}{AB} + \frac{A\left(R'_A + 3R'_x\right)}{6B} + \frac{R''_x}{2B} \qquad (3.48)$$

Disregarding the capacitive term and recalling that $AB = n$, the ESR value of the channel circuit is obtained:

$$ESR_{ch} = \frac{R_{ch} + R'_B}{n} + \frac{A\left(R'_A + 3R'_x\right)}{6B} + \frac{R''_x}{2B} \qquad (3.49)$$

Finally, the ESR of the whole structure of the proposed MOS capacitor is the addition of those corresponding to the gate and channel circuits:

$$\begin{aligned} ESR_T = &\frac{R_{ch} + R'_B}{n} + \frac{A\left(R'_A + 3R'_x\right)}{6B} + \frac{R''_x}{2B} \\ &+ \frac{R_G + R_B}{n} + \frac{BR_A}{6A} \end{aligned} \qquad (3.50)$$

As shown in expression (3.50), the total ESR depends on the matrix aspect ratio (A and B) and number of cells n. A more suitable arrangement shows better the dependencies on the matrix dimensions and that expression becomes as follows:

$$ESR_T = \frac{R_{ch} + R'_B + R_G + R_B}{n} + \frac{R'_A + 3R'_x}{6} \frac{A}{B} + \frac{R_A}{6} \frac{B}{A} + \frac{R''_x}{2B} \qquad (3.51)$$

Hence, given a number of cells n there will be a constant value of the ESR_T (the first term of (3.50)), plus a value depending on the matrix shape. According to this, the matrix shape dependent terms are grouped.

$$ESR_T = \frac{R_{ch} + R'_B + R_G + R_B}{n} + ESR'_T \qquad (3.52)$$

$$ESR'_T = \frac{R'_A + 3R'_x}{6} \frac{1}{r} + \frac{R_A}{6} r + \frac{R''_x}{2B} \qquad (3.53)$$

where r is the aspect ratio of the matrix, defined as $r = \frac{B}{A}$.

If the assumption that the last of term of (3.53) is negligible is made (which will be validated afterwards), it can be noted that ESR'_T presents a directly and an inversely proportional terms. Thus, there's an optimum value of ESR_T as a function of B/A.

The optimum matrix aspect ratio $r = B/A$, that provides the lowest value of ESR_T, is provided in (3.54).

$$r_{opt} = \sqrt{\frac{R'_A + 3R'_x}{R_A}} \tag{3.54}$$

And the corresponding minimum value of ESR'_T is:

$$ESR'_{T\_min} = \frac{\sqrt{R_A\left(R'_A + 3R'_x\right)}}{3} \tag{3.55}$$

Therefore, the value of the ESR of the optimum shaped structure is given by the following expression:

$$ESR_{T\_min} = \frac{R_{ch} + R'_B + R_G + R_B}{n} + \frac{\sqrt{R_A\left(R'_A + 3R'_x\right)}}{3} \tag{3.56}$$

which depends on the number of cells n, the dimension of each cell (W and L) and the technological parameters.

In the following, the previous assumption that the last term of (3.53) is negligible, will be validated.

From expression (3.51) it can be seen that when B takes high values, the third term (apart from the one independent of the matrix aspect ratio) is the most significant, and thus the assumption is valid. In case of low values of B, the validity of the assumption will depend on the following inequation:

$$\frac{R''_x}{2B} \ll \frac{R'_A + 3R'_x}{6}\frac{A}{B} \tag{3.57}$$

Which becomes:

$$A \gg \frac{3R'_x}{R'_A + 3R'_x} \tag{3.58}$$

The inequation (3.58) generally holds since the values of resistances are of a similar order and, for the case of B having a low value, A will have a high one (specially if the number of cells is high).

Figure 3.29 shows the error committed when evaluating the total ESR (in % over the total ESR) in case of ignoring the last term of (3.53), as a function of n and for different values of L (and the corresponding values of W) and for optimum

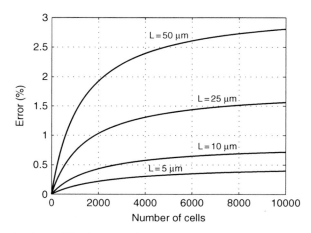

Fig. 3.29 Committed error if the last term of (3.53) is not considered in the matrix optimum aspect ratio calculation. Used technical values appear in the Table 3.2

shaped structures, according to expression (3.54). The error committed is evaluated as follows:

$$Error_{ESR}(\%) = \frac{ESR_{T\_C} - ESR_T}{ESR_{T\_C}} 100 \tag{3.59}$$

where ESR_{T_C} is the ESR with all the terms that appear in expression (3.51), and ESR_T is the same value without the $R_x''/2B$ term.

The results of Fig. 3.29 show that the error committed increases with the number of cells since the relative weight of the ignored term in (3.53) is more significant. The error increases as single cells become larger, although in any case it is negligible. Therefore, the assumption is validated for reasonable values of n and cell dimensions. The technological values used to calculate the results from Fig. 3.29 are presented in Table 3.2.

From expressions (3.24) and (3.56), it is observed that the total area and ESR depend on the number of parallel connected MOSCAPs. This is, although capacitive area is determined by required capacitance and the C_{ox} (from the particular technological process), the routing area depends on the single cell dimensions and their number.

As a consequence, several different configurations (in terms of number of cells and their dimensions) would provide the desired capacitance but different ESR and total occupied area values. Thus, a merit figure is defined to select the design that minimizes the product between the total occupied area and ESR, this is, that maximizes the following expression:

$$\Gamma_C = \frac{1}{ESR_T(A_{Cc} + A_{Cr})} \tag{3.60}$$

Table 3.2 Technological values used in example MOSCAP calculations

Technical parameter	Value	Technical parameter	Value
V_{gs}	1 V	C_{ox}	6.27 fF/μm^2
μ_N	376 cm^2/(sV)	V_{term}	25 mV
α	1/12	γ	12
a	0.8 μm	b	0.8 μm
R_{poly}	2.5 Ω	R_{MET1}	53 mΩ
R_{MET2}	53 mΩ	R_{MET3}	53 mΩ
δ_{cont}	1/0.72 cont./μm	δ_{via1}	1/0.76 Via1/μm
δ_{via2}	1/0.76 Via2/μm	R_{cont}	5 Ω/cont.
R_{via1}	3.5 Ω/via1	R_{via2}	3.5 Ω/via2

Characteristic	Value
Single cell optimum aspect ratio	44.51
Single cell minimum ESR	18.54 Ω

Figure 3.30 presents the results from the design of a 15 nF MOSCAP and 1 V of applied voltage. Total occupied area and capacitive area (A_{Cc}) are depicted as a function of the number of cells (Fig. 3.30a), as well as the total ESR (Fig. 3.30b)

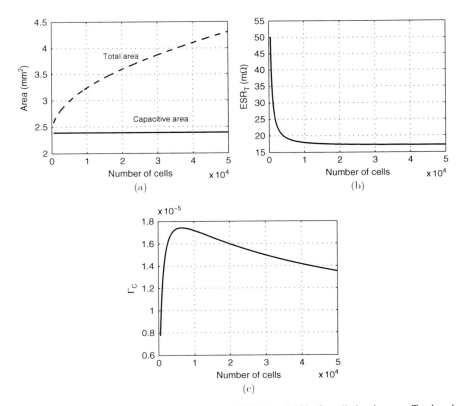

Fig. 3.30 Design example results of a 15 nF MOSCAP and 1 V of applied voltage: **a** Total and capacitive area, **b** Total ESR and **c** merit figure

and the resulting merit figure (Fig. 3.30c). In this case, the selected design would be: $n = 6,593$, $L = 2.85\,\mu m$, $W = 127.02\,\mu m$, $ESR_T = 18.6\,m\Omega$, $A_C = 3.08\,mm^2$, $A \approx 320$ and $B \approx 21$. It is interesting to observe that an important amount of occupied area is spent in the MOSCAPs interconnections (in the figure, this is the difference between both lines).

3.2.3 Comparison with Poly-Poly and Metal-Metal Capacitors

The two most common methods for implementing an on-chip capacitor in standard CMOS consider using either two polysilicon layers or two metal layers as electrodes of the capacitor (MIM capacitor, which stands for *Metal-Insulator-Metal* capacitor), yielding voltage-independent capacitors. Although they are actually special technological modules for CMOS technologies, they are present in most of the mixed-signal standard processes, and as such, in this work they are considered as standard CMOS-compatible.

Both kinds of on-chip capacitors should be compared to a MOSFET-based capacitor as regards to ESR, despite their relatively low specific capacitive density. According to the resistive component given by (3.15), the ESR is the addition of the resistance of each plate.

$$ESR = \left(R_{(MET/poly)1} + R_{(MET/poly)2} \right) \left(\alpha \frac{W}{L} + \beta \right) \tag{3.61}$$

being $\alpha = 1/12$ in case of a two-sided connection.

If no external resistance is considered, the ESR of a *poly-poly* or *metal-metal* capacitor has no minimum and can be theoretically reduced to negligible values by means of an extremely low W/L ratio, which becomes impractical for realistic implementations. Figure 3.31 shows the ESR comparison of a single-transistor MOSCAP as compared with these other implementations cases, as a function of W/L.

If an optimum single-transistor MOSCAP is compared with *poly-poly* or *metal-metal* capacitor with the same area and W/L ratio, it can be obtained the frequency band where each capacitor offers the lowest RC impedance. In (3.62), $f_{1\to 2}$ is the frequency value where both impedances present the same value.

$$f_{1\to 2} = \frac{1}{2\pi\, A C_1 C_2} \sqrt{\frac{C_1^2 - C_2^2}{ESR_1^2 - ESR_2^2}} \tag{3.62}$$

where 1 and 2 subindexes indicate the different technical implementations for the ESR values of capacitors with the same W/L ratio, and their capacitive densities. This is illustrated in Fig. 3.32, for technological parameters from Table 3.2, a 1 mm^2 area capacitor and $W/L = 44.51$. From that figure it can be noted that, in general, the MOSCAP is the better choice at low and medium frequencies because of its higher capacitive density. At higher frequencies (over 1 GHz for practical

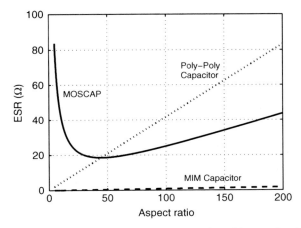

Fig. 3.31 Single cell ESR comparison of 3 different kinds of on-chip capacitor implementations, as a function of their aspect ratio

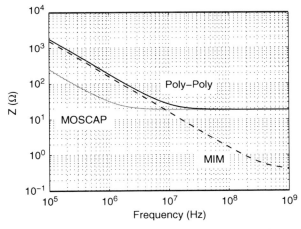

Fig. 3.32 Single cell impedance comparison of 3 different kinds of on-chip capacitor implementations, as a function of frequency. All of them occupy 1 mm² and present an aspect ratio of 44.51

sizes), where the impedance value is dominated by the ESR value, the *metal-metal* capacitor is a better choice because of its extremely low ESR. In this example, $C_{MIM} = 1\,\text{fF}/\mu\text{m}^2$ and $C_{poly\_poly} = 0.86\,\text{fF}/\mu\text{m}^2$ were taken as capacitive densities for MIM capacitor and *poly-poly* capacitor, respectively.

At this point, another expected benefit from the proposed matrix structure should be considered. This is due to connections by means of 3 bottom level metal layers, almost throughout the whole capacitor area, that results in a *sandwich structure* of *metal-to-metal* capacitors (which is a well known technique to improve the capacitance per area of on-chip capacitors). Although this 'MIM' capacitors have not been explicitly developed for this purpose (and therefore provide low capacitance per area), they are parallel connected to MOSCAP, adding an extra capacitance. Consequently, a reduction of the resulting impedance at higher frequencies is expected, since they provide very low ESR because of the metal plates.

In Fig. 3.33, an example of the extra impedance reduction due to the opportunistic sandwich structure is depicted. In this example case, the metal plates of the *sandwich structure* were considered to present the same ESR as the MIM capacitor of Fig. 3.32 ($R_\square = 53\,\text{m}\Omega$), but its capacitive density is lower ($C_{mm} = 36.3\,\text{aF}/\mu\text{m}^2$). Since parasitic capacitance between *metal1*, *metal2* and *metal3* is considered its plate area is twice the capacitor area. Furthermore, if a more accurate model is desired, an additional parasitic capacitor between the gate polysilicon and the *metal1* layer could be included, taking into account its corresponding ESR. Even more important than this is to consider that at so high frequency operation, the parasitic inductance should be modeled and taken into account (since it strongly determines the impedance behavior).

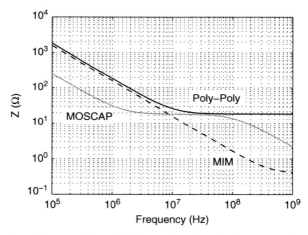

Fig. 3.33 Inclusion of the parasitic sandwich capacitor from the interconnection structure to the impedance comparison of 3 different kinds of on-chip capacitor implementations, as a function of frequency. All of them occupy 1 mm² and present an aspect ratio of 44.51

3.3 Tapered Buffer Model

3.3.1 State of the Art

Tapered buffers have been historically developed and analyzed in the literature, mainly in the field of drivers for digital IC pads. In 1975 Lin and Linholm [86] proposed a delay model based on the propagation delay between two minimum inverters and Jaeger [87] found the optimum tapering factor that minimized the total propagation delay. Afterwards, Nemes [88] added the concept of parasitic delay of an inverter without load. Subsequently, Li et al. [89] used the *split-capacitor* model to determine the total propagation delay of a tapered buffer. In 1991, Sutherland [90] introduced the delay estimation by means of the logical effort concept.

Although all these works deeply analyzed the propagation delay of tapered buffers, energy consumption is of paramount interest in high frequency switching power converters application. Fewer works have studied the power consumption of these circuits. Choi and Lee [91] proposed a capacitive approach based on the *split-capacitor* model. However, the capacitive approach does not account for the short-circuit current, which becomes more important as the transistor channel length is shrunk down in most recent technologies. Cherkauer and Friedman [92] added to the capacitive approach the short-circuit current estimation based on the Sakurai short-channel α-power model of transistor operation [93]. Unfortunately, nowadays α-power model parameters are unusual and require non obvious characterization to derive them (from test structures measurement or simulation).

More recently, contributions about tapered buffer design focused on their application to switching power converters have been reported (Stratakos et al. [94], Kursun et al. [35]) proposing tapering factor values that aim to reduce the corresponding switching losses. In this respect, an interesting work from Takayama based on the *split-capacitor* idea is pointed out [95].

As it will be justified afterwards, three different parameters from the tapered buffer performance need to be modeled: its power consumption, the *fall-rise* time at its output voltage (since power MOSFETs switching losses depend on it) and the input-to-output signal propagation delay of the overall chain of inverters.

As mentioned before, the last of these items (signal propagation delay) was well modeled by Nemes [88]. Thus, just the results are reminded here since they will be necessary. According to this model, the total signal propagation delay of a tapered buffer can be estimated from the performance of a single inverter, having its transistors minimum dimensions. From a very simple transient simulation the *intrinsic delay* t_{di} of an unloaded single inverter can be determined. Having loaded the minimum inverter with an identic one at its output, the measured extra delay is the so-called *unitary-effort delay* t_{de1}. Then, the overall delay of a chain of inverters (each of them scaled from the previous one by a tapering factor f) can be determined.

$$t_d = n(t_{di} + f t_{de1}) \qquad (3.63)$$

where t_d is the total propagation delay, n is the number of inverters and f is tapering factor between 2 consecutive inverters.

3.3.2 Proposed Energy Consumption Model

The proposed energy consumption model is based on the amount of charge required to change the state of a digital inverter. Charge is observed rather than current because for switching losses evaluation it is more convenient to model the energy (that depends on voltage and charge) spent in each switching cycle than to model the power, since the power changes with switching frequency.

Both *intrinsic charge* and *unitary effort charge* concepts (Fig. 3.34) are used to determine the total charge spent in a switching cycle by the i-th stage of the inverter chain. As observed in Fig. 3.34, both magnitudes can be very easily measured.

Intrinsic charge Q_i. Charge spent due to current flowing from V_{dd} in a unitary inverter during a state transition.

Unitary effort charge Q_{e1}. Charge spent due to the additional current flowing from V_{dd} during the state transition, when a unitary inverter is loaded with an identical inverter.

$$\text{Switching charge} = f^{i-1} Q_i + f^i Q_{e1} \tag{3.64}$$

In expression (3.64), f represents the tapering factor, and i is the exponent corresponding to the i-th stage, being the first one $i = 1$. The total charge per cycle supplied from the voltage source to the whole n-inverters chain is:

$$Q_T = Q_1 + Q_2 + \ldots + Q_n = \left(\frac{Q_i}{f} + Q_{e1}\right) \sum_{i=1}^{n} f^i = (Q_i + f Q_{e1}) \frac{f^n - 1}{f - 1} \tag{3.65}$$

And consequently the energy consumption is given by:

$$E_{driver} = V_{dd} (Q_i + f Q_{e1}) \frac{f^n - 1}{f - 1} \tag{3.66}$$

The most common definition of the tapering factor [88] is found as the ratio between the equivalent input capacitance of the first stage (C_{in}), and the equivalent

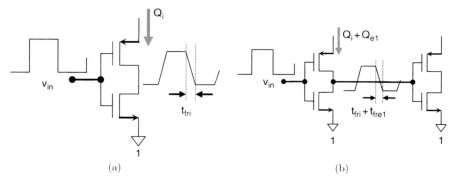

(a)   (b)

Fig. 3.34 Main concepts to estimate both energy consumption and *fall-rise* time of a tapered buffer: **a** intrinsic charge and fall-rise time; **b** unitary effort charge and fall-rise time

rather constant ratio between the model results and the simulation measurements exists.

As shown in Fig. 3.38, the t_{fr} increases as the number of inverters is reduced (and the corresponding tapering factor increases). This trend becomes clearer in the transient simulations of Fig. 3.36, where falling and rising edges of the output voltage are presented for several inverters chains (with the corresponding different tapering factors), keeping the power MOSFET size constant.

Again, more accurate results are obtained from the model if the t_{fr} calculation is carried out with the tapering factor definition from the ratio between charges (3.69) rather than from the transistor channel width, as shown in Fig. 3.39.

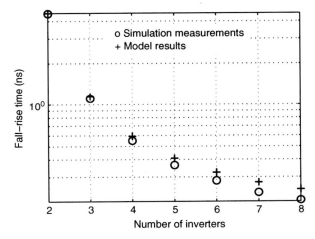

Fig. 3.39 Fall-rise time as a function of the number of stages, with the tapering factor definition from (3.69)

3.3.4 Optimized Design of Tapered Buffer Driving a Power MOSFET

The identification of two opposed trends in the dependence of energy consumption and fall-rise time on the number of inverters of the tapered buffer, suggests the existence of an optimized design of the tapered buffer as regards energy consumption (this is, when aiming to minimize the overall switching losses encompassing tapered buffer and power transistor).

To find this optimized design, a relationship between fall-rise time and switching losses in the power transistor is required. As it will be shown afterwards, in this work a linear relationship between a fraction of the switching energy losses of a power MOSFET and the driver output fall-rise time, is proposed.

$$E_{TRT\_D} = t_{fr} b_D \tag{3.71}$$

where b_D is the linear relationship coefficient between the t_{fr} value and a fraction of the power MOSFET switching losses, and is expected to depend on the power transistor channel width and the switching conditions (i.e. switching current and voltage).

Resorting to the modeling provided by Eqs. (3.71) and (3.66), an expression for the overall power losses related to power driver design, can be obtained:

$$E_D = E_{TRT\_D} + E_{driver} = t_{fr}b_D + V_{dd}\left(Q_i + f Q_{el}\right)\frac{f^n - 1}{f - 1} \tag{3.72}$$

The last step is to substitute the tapering factor by its expression in terms of number of inverters n (as given in (3.67)). The number of inverters is preferred as a design variable because it is constrained to natural values, whereas f varies continuously. Finally, the optimized design of the tapered buffer that not only minimizes its own losses, but minimizes the power transistor switching losses hence achieving the global goal is obtained as:

$$n_{opt} = \frac{\log(a_D)}{\log\left[\sqrt{\frac{V_{bat}(a_D-1)(Q_i+Q_{el})}{t_{fre1}b_D}} + 1\right]} \tag{3.73}$$

where a_D is the ratio used to define the tapering factor (e.g. $a_D = \frac{W_{power\_MOS}}{W_p+W_n}$).

The result of expression (3.73) needs to be rounded to the nearest natural value, since it represents the number of inverters of the tapered buffer. It should be noted that (3.73) could provide both even and odd values. Therefore, the possible sign inversion in the control signal transmission should be taken into account in the design phase.

Figure 3.40 depicts the total losses related to the driver design (which implies the driver energy consumption as well as the power transistor losses due to a non-zero fall-rise time, as exposed in (3.72)), as a function of the driver number of stages. In this case, the selected design would be composed of 3 inverters, being the tapering factor defined as the ratio between the transistor channel widths. Consequently, the previous observed error in the energy losses estimation is still present, although the model still allows to identify the design that minimizes the total losses. In the example, the relationship between the t_{fr} and the power transistor switching losses was determined to be $b_D = 0.06264$ J/s, from transistor-level simulations.

At this point, attention should be paid on an additional degree of freedom in the driver design. This is the PMOS channel width of the minimum inverter (the first driver stage). A conventional design would impose the minimum channel width for the NMOS transistor, and the channel width for the PMOS transistor that equalizes their respective conductivity. However, this could result in higher power consumption or greater fall-rise time. Thus, it is interesting to choose the ratio W_p/W_n (or the value of W_p, provided that the NMOS transistor presents the minimum channel width) that minimizes the overall energy losses. In this sense, it is interesting to run a batch of simulations and the corresponding measurements to observe the evolu-

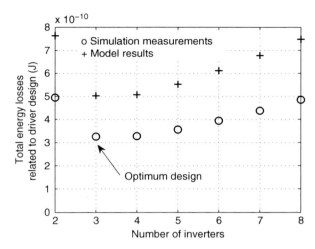

Fig. 3.40 Total losses as a function of the number of inverters. The optimum design that minimizes the overall losses is identified

tion of the four required parameters (Q_i, Q_{e1}, t_{fri}, t_{fre1}) as a function of the PMOS transistor channel width (Fig. 3.41).

Then, expression (3.73) should be evaluated for any W_p value (and the corresponding parameters values) to obtain the design that provides the minimum power losses. With all this information, the evolution of the resulting energy losses from the accordingly optimized designs is presented in Fig. 3.42.

The results from Fig. 3.42 could be somewhat surprising, since total losses are directly related to charge parameters as well as fall-rise time, as stated by (3.72). Because of the simulation measurements from Fig. 3.41 that increase for higher W_p values, a minimum in the total losses curve would be expected. Nevertheless, it must be taken into account that the optimum tapering factor decreases as W_p increases

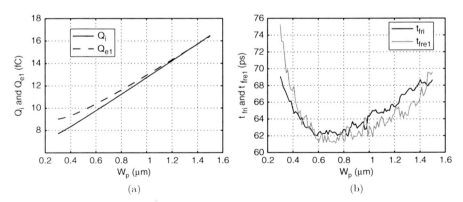

Fig. 3.41 a Q_i and Q_{e1} variation as a function of the PMOS channel width of the minimum inverter. **b** Similar essay for t_{fri} and t_{fre1} parameters. In both cases, $W_n = 0.3\,\mu m$

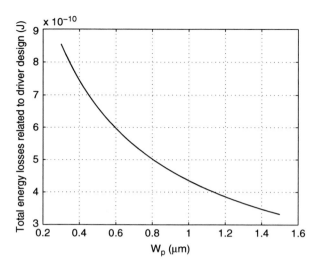

Fig. 3.42 Total losses as a function of the minimum inverter PMOS transistor channel width

(Fig. 3.43), since the ratio between the output power transistor gate capacitance, and the input capacitance of the minimum inverter is reduced.

Ideally, it is observed that higher W_p would reduce the total power losses related to the driver design. However, this might be a misleading reasoning because it must be considered that the input signal of the tapered buffer is driven by the control circuitry. As a consequence, the minimum driver input capacitance should not be increased more than the output driving capability of the control circuitry.

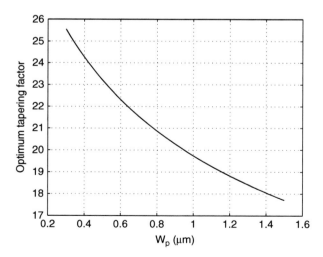

Fig. 3.43 Optimum tapering factor (that minimizes power losses) as function of the W_p value

3.3.5 Propagation Delay t_d Constraint in the Driver Design

In the previously presented design procedure no timing constraints have been considered. In this sense, Fig. 3.44 represents the fall-rise time that results from the optimization of the tapered design (according to (3.73)). Here, additional considerations should be done regarding the operating switching frequency of the converter to be designed; and some constraints should be applied if the ratio between the duration of the any of the control signal states, and the fall-rise time is too small.

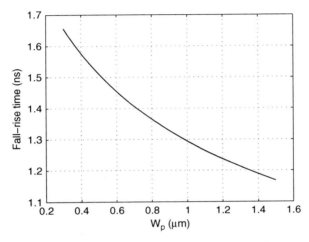

Fig. 3.44 Resulting t_{fr} when the tapered buffer design is optimized for any W_p value

Furthermore, overall malfunction could result if propagation delay (t_d) is sufficiently close to the control signal states duration. Specially, in the case where any power transistor requires its own driver; and they are supposed to exhibit equal delays when designing the control scheme. To prevent these problems, signal propagation delay is evaluated for any driver, by means of the well known expression (3.63).

Figure 3.45 shows the variation of the *intrinsic delay* t_{di} and the *unitary effort delay* t_{de1}, required to calculate the propagation delay.

The information from Fig. 3.45 is used to evaluate the total signal propagation delay as a function of n and W_p, as shown in Fig. 3.46. From these results, the propagation delay information can be used to constrain the tapered buffer design space. As an example, in Fig. 3.47 the total energy consumption (from the tapered buffer itself plus the power MOS switching losses) is exposed as a function of W_p and n. In this figure, the projection of the design area constrained to those designs that produce a propagation delay lower than 1.15 ns is marked with a white contour. As a consequence, the desired design configuration is the one that generates the lowest energy losses, while keeping a propagation delay lower that 1.15 ns. In this

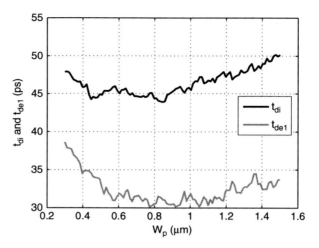

Fig. 3.45 t_{di} and t_{de1} parameters variation as a function of W_p

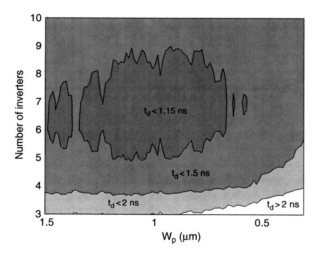

Fig. 3.46 Total signal propagation delay t_d as a function of the number of inverters n and the minimum inverter PMOS channel width W_p

case, this optimized design corresponds to $n = 5$ and $W_p = 1.19\,\mu m$ (marked with a white cross on the figure).

The noisy appearance of Figs. 3.41b, 3.45 and 3.47 is mainly due to signal measurement inaccuracies on the simulation results.

At this point, it should be observed that, as regards the optimum driver design, only energy losses are minimized. The reason for this can be found in the design results of Chap. 2, specially in Fig. 2.24b, where it is observed that the driver area occupancy becomes insignificant in front of the area occupied by the reactive

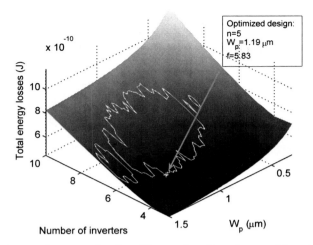

Fig. 3.47 Total energy losses as a function of the number of inverters n and the minimum inverter PMOS channel width W_p. The *marked area* includes all the designs that produce a propagation delay lower than 1.15 ns. The selected design that minimizes the energy losses and holds $t_d \leq 1.15$ ns is marked with a *white cross*

components, whereas the driver power losses have a great impact on the overall power efficiency.

The results of the design optimization (Fig. 3.47) yield that the optimum tapering factor ($f = 5.83$) is greater than the one used in typical digital applications that maximize speed (in terms of propagation delay), whereas it is not so small to just minimize the driver energy losses. This arises from the fact that a joined codesign of the tapered buffer and the power MOS transistor (afterwards presented) is proposed in this work, from the modeling of both the driver energy consumption and output voltage fall-rise time (t_{fr}) (assuming a relationship between this and some of the power MOS switching losses). Being this models similar to the one previously existing for the propagation delay [88], and very simple to obtain from transistor level transient simulations.

Additionally, the exploration of the impact of the W_p/W_n ratio (or more directly, W_p width) as a degree of freedom of the design procedure, leads to a driver design improvement (as regards total energy consumption).

Finally, the propagation delay has been presented as a constraint of the driver design space, in order to guarantee the convenient driver operation, by means of assuring that it is not higher than the maximum tolerable by the control scheme.

3.3.6 Area Occupancy Considerations

Although only energy losses are minimized by means of the optimized driver design, it is interesting to compute the corresponding occupied silicon area in order to include it in the total design space exploration. Thus, it is calculated as the addition of the area occupied by all the driver transistors considering not only the gate area, but also the sources and drains diffusion areas.

$$A_{driver} = (W_p + W_n)(L_{min} + L_{diff}) \frac{a_D - 1}{f_{opt} - 1} \qquad (3.74)$$

where L_{diff} is the drain and source diffusion length, extending aside of the transistor gate.

3.4 Power MOSFET Losses Model

3.4.1 State of the Art

Historically, the high-power low-frequency application of switching power converters required power switches with a very low on-resistance (R_{on}), and high-current driving capabilities. However, the miniaturization requirements that stems from its application to portable devices pushes the trend to continuously increase the switching frequency. This results in the fact that switching losses start to become as important as the conduction losses.

Therefore, it is very interesting to appropriately model both conduction and switching losses in power switches (that, in case of standard CMOS implementation are MOSFET transistors), in order to foresee the energy losses and estimate the overall efficiency of the power converter.

Following the design lines proposed in this work, the selected power loss model should be inserted in the overall design space exploration scheme. Hence, this model is required to account for variations of the electrical switching conditions (i.e. switching voltage and current, and gate driving voltage), switching transition duration, and it should be suitable for both DCM and CCM operating modes. Even more important than this is that it must be parametrizable by the power transistors channel width (provided that they present minimum channel length), since in microelectronic design it is the major degree of freedom and advantage.

Because of their simplicity, *conduction losses* are relatively easy to model. Commonly, they are computed as the product between the square of the transistor current RMS value and its equivalent *on-resistance* (R_{on}). The main handicap is to determine the R_{on} value accurately enough. Its most habitual expression is obtained from the *quadratic-model* of the MOSFET transistor (and the corresponding technical parameters), considering its operation in the triode region and a very low drain-to-source voltage drop [96] (although already presented in Sect. 2.2.2.3, it is reprinted here just as a reminder):

$$R_{on} = \frac{L_{ch}}{\mu_{N/P} C_{ox} W_{power\_MOS}(V_{gs} - V_{TN/P})} \qquad (3.75)$$

Nevertheless, in case of deep-submicron technologies the quadratic-model could become no longer accurate and a different expressions might be required (e.g. from the α-power model [93]). On the other hand, Kursun made use of some simulation results to extrapolate the R_{on} value as a function of the channel width, for a given gate-to-source voltage [34, 35].

As regards *switching losses*, the previously developed work is much more heterogeneous. This is because they are more difficult to model due to the high number of non-linear parasitic components that show their effects during the transistor state transition, as well as the effect of inductive clamped switchings.

The most basic models for switching energy losses evaluation just consider the integration of the $v_{ds}i_{ds}$ product during the switching interval; where different variations may include linear (as the model presented in Sect. 2.2.2.3) or more realistic waveforms. A further step in the model complexity has been carried out by several authors, specially in the field of discrete power transistors for switch-mode power supplies [97–101]. These are somehow based on the reproduction and analysis of the internal transistor currents and voltages waveforms. These works can include the non-linear capacitors as well as the parasitic inductors from the package pins. Unfortunately, these models generally rely on the knowledge of several component parameters (such as the gate resistor value) which are usually provided by the manufacturer (in the datasheet), but result very difficult to model in the integrated framework. To justify this assertion, note that the equivalent gate resistor of an integrated power MOSFET depends on the driver design and layout, the power MOSFET layout, and the drain and source diffusion resistivity (which may result non-linear). However, Brown [99] preferred to use the equivalent charge invested to change the voltage of some of the parasitic capacitors, in case that the timing of the different phases that happen along the switching interval could not be easily determined. A very interesting advantage of these models is that some of them provide an analytical calculation of the switching losses that could yield an optimized design, by means of analytical methods.

Another different approach to switching losses evaluation (more focused on the on-chip integration) claims that the energy lost in switching transitions is due to charging and discharging the parasitic capacitances of the MOSFET transistor, specially the *x-node* capacitor (in case of a buck converter) [34, 35, 96, 102]. Although this approach yields very simple expressions, suitable for an analytical optimization of the transistor design, it does not consider any dependency of the switching losses either on the switching interval duration, nor on the inductor current at the switching instant. Additionally, the presence of an inductive clumped load connected at the switching node, is not taken into account in most of these works. Consequently, no switching loss reduction due to soft-switching techniques such as, *Zero-Voltage-Switching* (ZVS) or *Zero-Current-Switching* (ZCS) is considered.

Finally, a very interesting switching losses model was proposed by Williams et al. in 1995 [103]. This approach, specifically developed for discrete vertical DMOS-FETs, takes into account the energy spent in charging and discharging the non-linear parasitic capacitors, but also includes the effect of discharging some of them by means of the inductor current (which is a lossless process), and thus, some energy is saved. This approach makes unnecessary to model the gate resistance, since no waveform is reproduced. Moreover, it also includes the voltage and current overlap during switching transition, which adds to the model the dependence of switching losses on the state change duration and the instantaneous inductor current. This interesting work even includes the conduction losses by an R_{on} expression similar to 3.75. As a consequence, optimum values for gate oxide thickness, gate driving voltage and die area, are discussed.

In order to appropriately design the power MOSFET to balance both conduction and switching losses, in the discrete components environment some authors define a merit figure to be minimized as the product $Q_{in}R_{on}$ (being Q_{in} the charge spent to drive the transistor gate [99]). This should yield the convenient technology or device selection to implement the power switches (since in the discrete components framework no specific design of the power switch is available). Nevertheless, in [104], Ajram and Salmer developed a fully integrated switching power converter on GaAs, and evaluated the suitability of different power switch options by means of the same merit figure.

In case of on-chip power MOSFET, and optimum channel width can be designed to minimize the total energy losses [105]. In this sense, proposed analytical models for switching and conduction models led to a closed expression of the optimum channel width W_{opt} [33–35, 96]. Furthermore, in [96], Musunuri proposed to dynamically adapt the power MOSFET channel width to maximize efficiency as the converter output power changes.

In this work, a model for the switching losses evaluation is proposed, which closely matches the one presented by Williams for vertical DMOSFET transistors [103]. It is intended to include the energy wasted to charge the parasitic capacitances of the power MOSFETs (but considering possible soft-switching transitions) avoiding the need to model the gate resistance for this purpose. In addition to this, the model accounts for the dependency on the inductor current during the switching interval, and its duration (relating the switching losses to the driver design, too). It is suggested to associate the energy losses to an specific state change of the whole converter, rather than to any individual transistor, leading to a concurrent design of all the power MOSFETs and their corresponding drivers. It is important to note that the loss model herein presented considers that optimum *dead-time* is applied to the switching scheme (which can be achieved by means of a dedicated control loop, as will be explained in another chapter). However, details on the body-diode conduction due to a suboptimum dead-time, and their corresponding switching losses, have been widely addressed in the literature [84, 97, 106].

3.4.2 Switching Losses Evaluation

The power MOSFETs switching losses evaluation is split in two different terms: energy losses due to the charge process of the parasitic capacitors, and energy losses due to a non-instantaneous transition from the on-state to the off-state. Since both loss mechanisms act simultaneously, the total switching losses are computed as the addition of both of them. In any case, energy is computed rather than power because the later depends on the switching frequency, which could be variable (depending on the applied control method).

$$E_{sw} = E_{sw\_C} + E_{sw\_R} \tag{3.76}$$

3.4.2.1   Capacitive Switching Losses (E_{sw_C})

First, any voltage change on any parasitic capacitor is evaluated, and the lost energy to achieve those voltage changes are computed. This computation is carried out for any change of the circuit state. Consequently, energy switching losses are associated to a determined state transition, rather to a singular power MOSFET. The reason for this association is that, in DCM operation, some state changes may require just one of the power transistors to switch, although the resulting losses may result from the design of both transistors.

To evaluate capacitive losses, this work supposes that no inductive effect is present in the power MOSFET structure. The result of such assumption is that all the energy losses are due to the Joule-effect any time that charge is transferred through the parasitic resistors to change the parasitic capacitors voltage. The main advantage of this approach is that the energy lost during a capacitor charge process (from a constant voltage source) is an state function, and as a consequence, it becomes independent of the transition duration as well as the path parasitic resistors values (that are usually difficult to model), even when they are very low. In fact, it just depends on the capacitor value, the voltage source that charges it (V_{in}) and the initial and final capacitor voltages (V_0 and V_f, respectively). The lost energy can be directly computed by means of the (3.77) expression (Appendix B).

$$E = (V_f - V_0)C \left(V_{in} - \frac{V_f + V_0}{2} \right) \tag{3.77}$$

Figure 3.48 shows the process efficiency as a function of the final-to-input voltage ratio (assuming that the initial capacitor voltage is $V_0 = 0\,V$). It is interesting to note that it is a very lossy process, since the minimum losses are only achieved when the capacitor is fully charged (this is, that its final voltage is the same as the voltage source), and still they are the 50% of the input energy. For further details on the energy lost during a capacitor charge process, please refer to the Appendix B.

It is well known that, under the low output ripple assumption, the main node that changes its voltage along a switching cycle is the node were the drains of both power MOSFETs are connected to the inductor (in this work it is called the *x-node*, and its voltage is v_x). Additionally, power transistors gate voltage is changed by their corresponding drivers to turn them on or off (Fig. 3.49).

For the capacitive losses evaluation, the following method is proposed:

1. Having identified the operating mode of the switching converter (DCM or CCM), all state transitions must be separately analyzed to identify all the voltage changes in the circuit nodes surrounding the power MOSFETs.

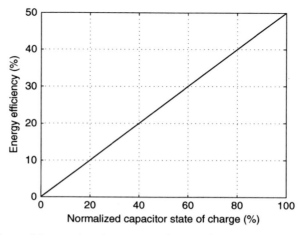

Fig. 3.48 Efficiency of the capacitor charge process from a voltage source as a function of the final state of capacitor charge (normalized to the source voltage)

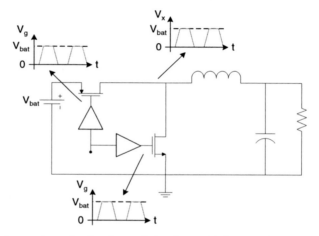

Fig. 3.49 Buck converter nodes that change their voltage with the state changes

2. Those state changes in which the voltage transitions are obtained by means of the inductor current are identified and excluded from the loss calculations, since this is a lossless process (if ideal dead-times are supposed, as well as very low parasitic resistances).

3. The non-linear behavior of the parasitic capacitors is included in the calculations by means of C-V tables, resulting from a bunch of simulation measurements. In the simulations, source and bulk terminals of PMOS and NMOS power transistors are connected to ground and battery voltage (V_{bat}), respectively. Then, voltage sweeps are run to capture the capacitor dependencies on the drain and gate voltages (thus, the results conform bi-dimensional tables). These tables are developed for any considered parasitic capacitor (junction capacitors, overlap

capacitors, etc...). In all the simulations, the transistor channel width is set to
$1,000\,\mu m$ and then, parasitic capacitor values are considered to be directly proportional to the channel width.

4. Since expression (3.77) can only be applied in case of a constant capacitor, a
staggered approximation is proposed to accumulate the overall energy losses
(Fig. 3.50). Then, the energy lost is computed considering a constant capacitor
charge between two adjacent voltage steps. In Fig. 3.50 the energy lost at the end
of any considered partial charge is depicted as a function of the capacitor value
(and the corresponding capacitor voltage). The total lost energy is computed as
the addition of all the partial losses corresponding to each subinterval.

$$E_{sw\_C} = \sum_{i=1}(V_{fi} - V_{0i})C_i \left(V_{in} - \frac{V_{fi} + V_{0i}}{2} \right) \tag{3.78}$$

where i identifies the corresponding magnitude of the considered ith subinterval,
and consequently $V_{0i+1} = V_{fi}$.

In the following, all state changes corresponding to a buck converter operation
will be studied. In Fig. 3.51 all the different states of a buck converter (T_{on}, T_{off}, T_i)
are depicted, and the existing voltage, roughly approached, at any switching nodes
is labeled by gray text. Additionally, gray boxes are used to mark the on-state of
each transistor, for any state.

It is important to note that for DCM operation the switching sequence is $T_{on} \rightarrow
T_{off} \rightarrow T_i \rightarrow T_{on} \rightarrow \dots$ This implies that turning-on of one of the transistors not
always implies to turn-off the complementary one (as happens in CCM operation).
Two of the state transitions are caused just by switching the state of one of the
transistors.

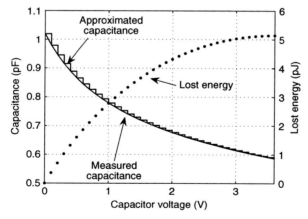

Fig. 3.50 Lost energy during the charge process of a non-linear capacitor (from a 3.6 V voltage
source), as a function of the capacitor voltage. The capacitor value is approached by an staggered
function that allows the use of (3.77) in any subinterval

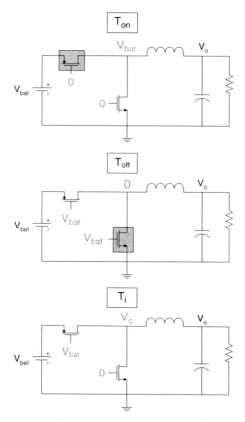

Fig. 3.51 Buck converter states. *Gray boxes* mark the on-state of the transistors, for any different state. *Gray text* is used to label the voltage value in those nodes that switch their voltage with the converter states

The voltage excursions of the power transistors gate-to-source voltages (v_{gsP} and v_{gsN}) as well as v_x for all the possible state transitions, are summarized in Table 3.4 for DCM operation. Furthermore, a gray row is used to mark that the $T_{on} \rightarrow T_{off}$ transition does not generate energy losses on the x-node capacitor, since it is discharged by means of the inductor current (provided that an optimum dead-time is applied between turning-off the PMOS and turning-on the NMOS). Table 3.5 exposes the same results if CCM operation is considered. In this table, the instantaneous currents values flowing through each transistor during the switching interval are also presented (which will be necessary afterwards to compute the resistive switching losses).

After the analysis of all voltage transitions, the parasitic capacitors that produce the switching capacitive losses must be identified and evaluated. Figure 3.52 shows all the parasitic capacitors connected to the x-node.

Table 3.4 Voltage excursions in the switching nodes of a DCM operated buck converter, and instantaneous transistor current value for each state change

	v_x	v_{gsP}	v_{gsN}	i_{PMOS}	i_{NMOS}
$T_{on} \to T_{off}$	$V_{bat} \to 0$	$V_{bat} \to 0$	$0 \to V_{bat}$	I_{L_max}	I_{L_max}
$T_{off} \to T_i$	$0 \to V_o$	$0 \to 0$	$V_{bat} \to 0$	0	0
$T_i \to T_{on}$	$V_o \to V_{bat}$	$0 \to V_{bat}$	$0 \to 0$	0	0

Table 3.5 Voltage excursions in the switching nodes of a CCM operated buck converter, and instantaneous transistor current value for each state change

	v_x	v_{gsP}	v_{gsN}	i_{PMOS}	i_{NMOS}
$T_{on} \to T_{off}$	$V_{bat} \to 0$	$V_{bat} \to 0$	$0 \to V_{bat}$	I_{L_max}	I_{L_max}
$T_{off} \to T_{on}$	$0 \to V_{bat}$	$0 \to V_{bat}$	$V_{bat} \to 0$	I_{L_min}	I_{L_min}

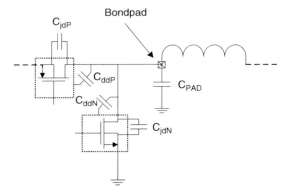

Fig. 3.52 Parasitic capacitors that contribute to the x-node total parasitic capacitance

Mainly, three different capacitors contribute to the total parasitic capacitor: C_{jd} accounts for the junction capacitors between the drain and the bulk of each transistor, whereas according to the BSIM3V3 MOS transistor model (which was used to measure the parasitic capacitors) C_{dd} represents the total capacitance observed from the drain terminal (apart from the junction capacitor). Obviously, both kinds of parasitic capacitors contribute to the total x-node parasitic capacitor, from both PMOS and NMOS transistors. Finally, the parasitic capacitor coupling to substrate corresponding to the bondpad (C_{PAD}), used to connect the inductor (according to the implementation proposed in Sect. 3.1.2), is also considered.

C_{PAD} is a constant capacitor whose value depends on the bondpad dimensions, the distance between it and the substrate and the $Si O_2$ dielectric permeability.

On the other hand, C_{dd} depends on the transistor dimensions and the voltage applied to drain and gate. No dependency has been taken into account on the source and bulk voltages, since these terminals remain constant along the whole switching

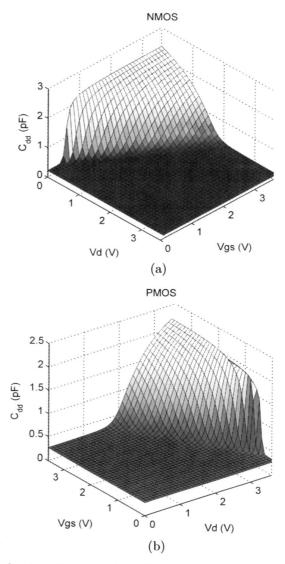

Fig. 3.53 C_{dd} as a function of the V_{gs} and V_d applied voltages. It is considered that the source and bulk terminals are connected either ground or V_{bat} for NMOS **a** or PMOS **b** transistors, respectively. 1,000 μm channel width has been considered

period. Figure 3.53 shows the variation of C_{dd} as a function of V_{gs} and V_d, considering that the bulk and the source terminals are connected to the ground or to the battery for the NMOS and PMOS, respectively. In this case, a 1,000 μm channel width has been considered. In spite of considering a 0.25 μm technological process, the selected transistors to implement the power switches present a channel length of

0.35 μm, since they are *input – output transistors*, which provide higher breakdown
voltage, that allow to work directly with the Li-Ion battery voltage (3.6 V).

As C_{jd} regards, it just depends on the transistor dimensions and the drain voltage, since the bulk voltage remains constant. Bi-dimensional tables have also been
developed in this case, because of computational needs (to better suit the computation algorithm). However, in order to provide a more clear view, just the results for
a single value of $V_{gs} = 1.8$ V (arbitrarily selected) are exposed. Again, $1,000$ μm
channel width transistors were the devices under test. The resulting junction capacitance for the PMOS and NMOS transistors is shown in Fig. 3.54.

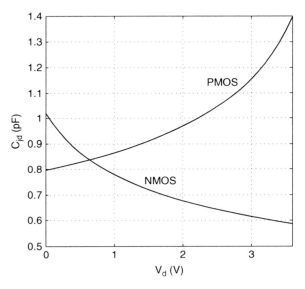

Fig. 3.54 C_{jd} as a function of the V_d applied voltage. Considering that the source and bulk terminals are connected either to ground or to V_{bat} for NMOS or PMOS transistors, respectively.
$1,000$ μm channel width has been considered

To sum up, according to Fig. 3.52 the total *x*-node capacitance (C_{V_x}) is the addition of all the parasitic capacitances.

$$C_{V_x} = C_{jdN}(V_{dN}) + C_{ddN}(V_{gsN}, V_{dN}) + C_{jdP}(V_{dP}) + C_{ddP}(V_{gsP}, V_{dP}) + C_{PAD} \quad (3.79)$$

From the previous expression (3.79), it should be observed the interdependencies
between the losses generated in one of the transistors, and the design of the other
one even in the case in which its state is not switched, as it happens in DCM operation. The reason for this is that the transistor that switches its state (and changes
v_x) must charge their own parasitic capacitances as well as those related with the
complementary transistor design.

DCM Operation

In the following, the energy lost in any of the state transitions when the converter is DCM operated, is evaluated. Figure 3.55 shows the temporal evolution of the x-node voltage, as well as the transistors gate-to-source voltages (please note, that v_{gs} is represented instead of the absolute value of the gate voltage), corresponding to the $T_{off} \rightarrow T_i$ transition. In this work, for the sake of simplicity, it is supposed that to change all three voltages takes the same amount of time, and that it happens simultaneously. Additionally, linear evolutions for the switching nodes voltage is supposed, too. Otherwise, it would be required to model the effective gate resistance.

From the voltages evolutions (Fig. 3.55a), the corresponding capacitor values are obtained (Fig. 3.55b), and finally the total capacitive losses corresponding to this transition can be computed (Fig. 3.55c). In this transition, the voltage source that charges C_{V_x} is the converter output capacitor (at the V_o voltage). Although it could seem that the charge is carried out through the inductor, in practice, the inductor will be short-circuited to avoid some noisy oscillations during the whole T_i interval (this will be deeply explained in a later chapter).

In Fig. 3.55, it is observed that the temporal axis does not show any value, this should be interpreted as that, for these energy losses evaluation purposes, no timing information is required (as explained in Appendix B).

The same analysis and considerations are carried out for the $T_i \rightarrow T_{on}$ transition, and the results are shown if Fig. 3.56.

As the $T_{on} \rightarrow T_{off}$ transition regards, it is observed that the x-node parasitic capacitor is discharged by means of the inductor current, and its charge is sent to the output capacitor. Provided that the total parasitic resistance is low, this is a lossless process. Thus, no capacitive losses are computed related to this transition. Additionally, it is noted that the energy losses corresponding to the transistors gates state change are computed and included in the driver loss model (presented in Sect. 3.3).

CCM Operation

The same reasoning applies for the case of a CCM operated buck converter. Consequently, only capacitive losses related to the $T_{off} \rightarrow T_{on}$ transition are computed. The corresponding results are presented in Fig. 3.57.

3.4.2.2 Resistive Switching Losses (E_{sw_R})

According to the proposed switching losses model, the resistive switching losses are due to a non-instantaneous change in the power MOSFET channel resistance. This is, it takes some time to pass the channel resistance from an infinite value (theoretically), down to the low on-resistance value, and vice versa. If inductor is modeled as a current source with a constant value along the whole switching interval (I_{Lsw}), this current is forced to pass through the time dependent resistor, and Joule-effect losses appear (Fig. 3.58).

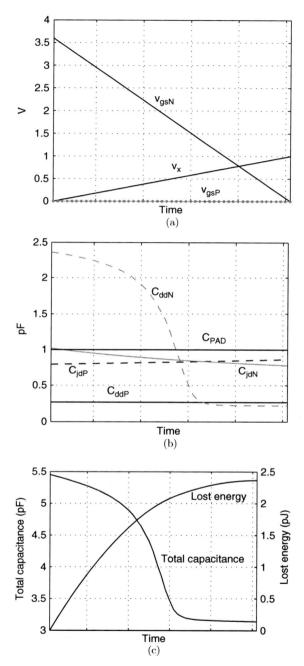

Fig. 3.55 Proposed approach to the temporal evolution of v_{gsP}, v_{gsN} and v_x **a**, and the corresponding parasitic capacitors evolution **b**, along the $T_{off} \rightarrow T_i$ transition (DCM). In **c**, the total capacitance evolution as well as the lost energy are shown

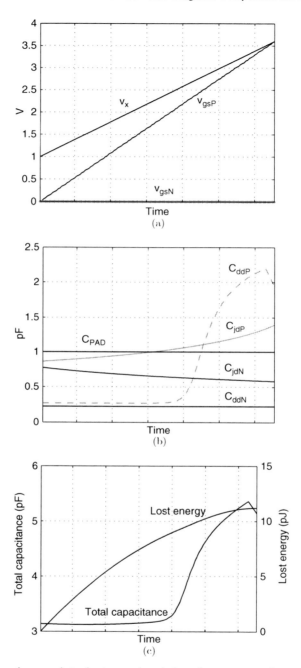

Fig. 3.56 Proposed approach to the temporal evolution of v_{gsP}, v_{gsN} and v_x **a**, and the corresponding parasitic capacitors evolution **b**, along the $T_i \rightarrow T_{on}$ transition (DCM). In **c**, the total capacitance evolution as well as the lost energy are shown

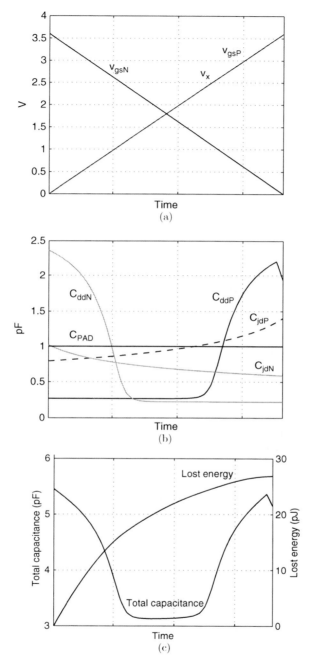

Fig. 3.57 Proposed approach to the temporal evolution of v_{gsP}, v_{gsN} and v_x **a**, and the corresponding parasitic capacitors evolution **b**, along the $T_{off} \rightarrow T_{on}$ transition (CCM). In **c**, the total capacitance evolution as well as the lost energy are shown

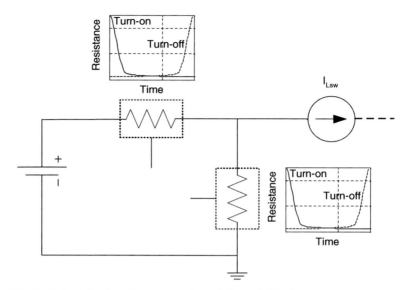

Fig. 3.58 Equivalent circuit used to evaluate the resistive switching losses

Basically, the resistive switching losses are obtained from resistance value, the switching interval duration ($t_{swon/off}$), and the inductor current at the switching moment.

$$E_{sw\_R} = I^2_{Lsw} R_{TRT} t_{sw} \qquad (3.80)$$

The first thing to determine is the switching interval duration, as resistive switching losses evaluation regards. In order to do that, transistor resistance as a function of the gate-to-source voltage is observed. Figure 3.59 shows the obtained transistor resistance as a function of the gate-to-source voltage (V_{gs}) and the transistor channel width (W_{power_MOS}), for a PMOS and a NMOS transistors (presenting both of them a 0.35 μm channel length, because of the aforementioned reasons). In the figure, axis resistance is presented in a logarithmic scale because of its great span (specially for low V_{gs} values). In addition to this, it must be mentioned that the resistance was obtained as the V_{ds}/I_{ds} ratio, being the drain connected to a 100 mA current source.

Since a linear evolution of the gate voltage has been considered, a direct relationship between the transistor resistance and the switching interval duration can be stablished. However, as observed in Fig. 3.59, very low V_{gs} values result in high resistance values, that would produce a voltage drop higher than the transistor drain-to-source voltage when it is turned off. Thus, to take into account such high resistance values becomes unrealistic.

As a consequence, in a turning-on action, the switching interval is considered to start when the gate-to-source voltage is the one that results in a channel resistance that, in turn, will produce a voltage drop equal to the corresponding voltage drop

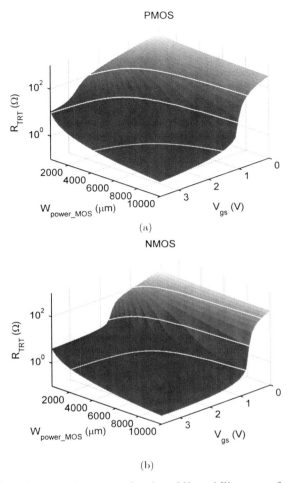

Fig. 3.59 Transistor channel resistance as a function of V_{gs} and W_{power_MOS} for a PMOS **a** and a NMOS **b** transistors. *White level lines* are used to mark the 1, 10 and 100 Ω values. The channel length was set to 0.35 μm

when the transistor is turned-off, given the inductor switching current (I_{Lsw}). If an arbitrary channel width is considered, this can be expressed as follows:

$$R_1 = R_{TRT}(V_{gs1}) = \frac{V_{ds\_off}}{I_{Lsw}} \tag{3.81}$$

In this sense, the turning-on action is considered to be ended when the channel resistance is equal to the transistor resistance when it is in its on-state (R_{on}). This means that the V_{gs} voltage has reached its final value, which in case of a Buck converter is equal to V_{bat}.

$$R_2 = R_{TRT}(V_{gs2}) = R_{on} = R_{TRT}(V_{bat}) \longrightarrow V_{gs2} = V_{bat} \qquad (3.82)$$

In (3.81) and (3.82), the V_{gs1} and V_{gs2} values are the values the produce the indicated resistance values.

Therefore, the switching interval duration (t_{sw}) is the time that it takes to change the transistor gate-to-source voltage from V_{gs1} to V_{gs2}, in case of a turning-on process. In the turning-off counterpart, the switching interval duration will take an analogous computation, because of the considered linear evolution of the V_{gs} voltage, and the voltage boundaries from V_{gs2} to V_{gs1}. However, no symmetric switching intervals can be considered (as turn-on and turn-off regards), since the switching inductor current will probably be different. Figure 3.60 is intended to clarify this reasoning.

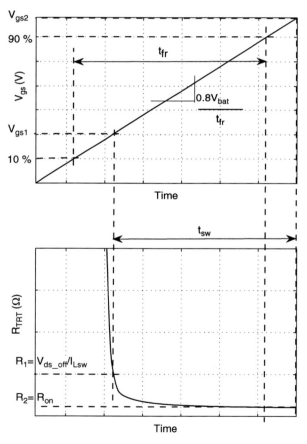

Fig. 3.60 t_{sw} duration evaluation from the R_1 and R_2 boundaries. It is related to the fall-rise time through the linear evolution of V_{gs}

From all the previous considerations, now the switching interval duration can be expressed as:

$$t_{sw} = \frac{V_{gs2} - V_{gs1}}{0.8V_{bat}} t_{fr} \qquad (3.83)$$

To obtain the transistor channel resistance value (R_{TRT}) as a function of the channel width (W_{power_MOS}) and the gate-to-source value (V_{gs}), a bunch of simulations were run and measured to build a look-up table (and the intermediate values are interpolated). Then, R_{TRT} is obtained as a function of time (through the different V_{gs} values). And finally, it is proposed to compute the average of all the different R_{on} values, and this value is at the end used to compute the resistive switching losses.

The inductor switching current for any of the transistor state changes, can be obtained from Tables 3.4 and 3.5, for DCM and CCM operation, respectively.

Thus, the total resistive switching losses for each transistor can be obtained from expression (3.84).

$$E_{sw\_R} = \left[\frac{V_{gs2} - V_{gs1}}{0.8V_{bat}} t_{fr} I_{Lsw}^2 < R_{on} > \right]_{turn-on} +$$
$$+ \left[\frac{V_{gs2} - V_{gs1}}{0.8V_{bat}} t_{fr} I_{Lsw}^2 < R_{on} > \right]_{turn-off} \qquad (3.84)$$

At this point, it should be reminded that in Sect. 3.3, the driver design was linked to the power MOSFET design by means of the b_D factor, which relates the driver output fall-rise time (t_{fr}) and a part of the power MOSFET switching losses (i.e. the switching resistive losses). Hence, (3.84) can be rearranged to determine this ratio, as follows:

$$E_{sw\_R} = b_D t_{fr} = t_{fr} \left[\left[\frac{V_{gs2} - V_{gs1}}{0.8V_{bat}} I_{Lsw}^2 < R_{on} > \right]_{turn-on} + \right.$$
$$\left. + \left[\frac{V_{gs2} - V_{gs1}}{0.8V_{bat}} I_{Lsw}^2 < R_{on} > \right]_{turn-off} \right] \qquad (3.85)$$

$$b_D = \left[\frac{V_{gs2} - V_{gs1}}{0.8V_{bat}} I_{Lsw}^2 < R_{on} > \right]_{turn-on} + \left[\frac{V_{gs2} - V_{gs1}}{0.8V_{bat}} I_{Lsw}^2 < R_{on} > \right]_{turn-off}$$
$$(3.86)$$

3.4.3 Conduction Losses Evaluation

As previously mentioned, conduction losses are straightforward to evaluate, provided that the effective channel resistance of the power MOS when it is turned-on (R_{on}) is well modeled.

$$P_{cond} = R_{on} I_{MOS}^2 \qquad (3.87)$$

The RMS value of the conducted current depends on many of the converter parameters (see Sect. 2.2.2.3) such as converter topology, output current, switching frequency, etc...

On the other hand, given a minimum channel length transistor, its R_{on} depends on some technical parameters, its channel width and the applied gate-to-source voltage. The main issue with expression 3.75 is that for short channel transistors, the model becomes inaccurate. Kursun et al. [34] avoided this issue by measuring the on-resistance (specified for a given gate voltage) for a single 1 μm transistor channel width (R_0), and then extrapolated the value for any channel width by dividing this value by the W_{power_MOS}.

Due to the resistive switching losses computation needs, a bunch of simulations and measurements were done to build up a table that accurately relates the R_{on} to the transistor channel width W_{power_MOS}, and gate-to-source voltage V_{gs} (although in the Buck converter case V_{gs} is considered to remain constant during the whole conduction state, and equal to V_{bat}, since it is the higher voltage of the circuit), which was already used for the resistive switching losses evaluation.

Figure 3.61 presents a comparison of the obtained results from expression (3.75), from the Kursun's expression and the data table, obtained from simulations measurements.

From these results it is observed that the classical expression becomes too optimistic and foresees a lower resistance than the actual, whereas Kursun's approach provides accurate results from just one simulation measurement.

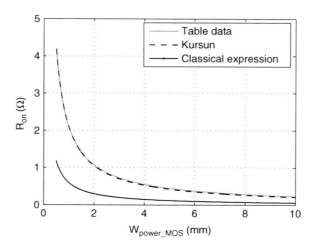

Fig. 3.61 R_{on} results comparison: (3.75) expression evaluation, Kursun's model and simulation measured data. $V_{gs} = 3.6$ V, $L_{ch} = 0.35$ μm, $\mu_N = 393$ cm^2/(Vs), $C_{ox} = 4.93$ fF/μm^2, $V_{TN} = 0.55$ V

3.4.4 Area Occupancy Considerations

Throughout this section, no area occupancy considerations have been presented. In this work, no special design strategy has been developed to reduce the occupied area of silicon. Thus, the total occupied area by the transistors is computed as a function of the transistors channel dimensions and the sources and drains diffusion areas (similarly as in the case of tapered buffers design).

$$A_{MOS} = W_{power\_MOS}(L_{min} + L_{diff}) \tag{3.88}$$

However, it would be very interesting to study and eventually develop special layout techniques that not only reduce the area occupancy, but also relatively reduce some of the undesired transistor parasitic effects (e.g. junction capacitances, driven current capability, etc...). In this sense, very interesting works have been developed by other authors [107–110].

3.4.5 Power MOSFETs and Drivers Codesign Procedure

In Sects. 3.3 and 3.4, models for the energy losses evaluation of tapered buffers and power MOSFETs have been presented. In the tapered buffers design procedure it was explained that it is linked to the power MOSFETs design through the b_D factor.

Hence, here it is proposed a procedure to carry out the joint design of both power transistors, as well as their corresponding drivers. This design procedure is focused to lost energy minimization, while no area considerations are taken into account, because the design of power MOSFETs and drivers strongly impacts on the overall efficiency, as observed in Chap. 2.

As previously mentioned in Sect. 3.4.2.1, the concurrent design of both power transistors is proposed in this work. As a consequence, the total lost energy evaluation is a function of two different variables (W_P and W_N), plus a set of external conditions imposed by the rest of the converter circuit (I_{L_min}, I_{L_max}, V_o,...) and technological parameters (L_{min}, C_{ox},...).

Additionally, many of the used expressions are non-linear and in many cases, look-up tables are used to obtained important parameters values (R_{on}, C_{dd},...).

Inevitably, no analytical methods could be used to optimize the design of the power MOSFETs (and their associated drivers), and numerical methods are used to optimize the design.

To clarify the whole process, first, the total energy losses evaluation is explained, and then, some discussion is presented regarding the total design optimization (to minimize the lost energy).

Figure 3.62 shows a diagram of the proposed evaluation procedure.

1. First of all, the converter operating mode must be determined, in order to properly identify all the possible converter states.

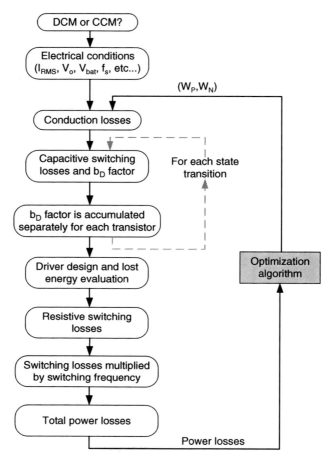

Fig. 3.62 Proposed procedure to concurrently design the power transistors as well as their associated drivers, so as to minimize their power losses

2. Then, all the relevant electrical conditions externally imposed by the converter operation (switching frequency, inductor, capacitor, control signal duty cycle, etc...), are used to fill the data from Tables 3.4 and 3.5. RMS values of the transistors currents are required, too.
3. Then, a pair of values (W_P, W_N) are proposed by the optimization algorithm and the conduction losses are firstly evaluated.
4. After that, the capacitive switching losses associated to any switching node (excluding transistors gates terminals) are computed, as well as the b_D factor for each transistor. This is carried out for all the converter state transitions. Please note that in case of *Zero-Current-Switching*, $b_D = 0$.
5. The different obtained b_D values are accumulated separately for each transistor.

6. With the power transistor channels widths and the corresponding b_D factors, the optimum driver is designed for any transistor and its energy losses are evaluated.
7. All the different switching losses are added and multiplied by the switching frequency (f_s).
8. The resulting switching power losses are added to the previously calculated conduction losses and reintroduced to the optimization algorithm.
9. The optimization algorithm recursively proposes (W_P, W_N) pairs, until the power losses minimization is ended.

Several solutions can be applied as optimization algorithm. In the following some of them are discussed.

- The most obvious way to determine the transistors channels widths that minimize the lost energy, is to explore the whole design space. Despite it assures the design optimization, it becomes very computing-intensive because of the large span and resolution of variables.
- *Geometric programming* is an analytical optimization method for non-linear systems [38]. It is based on the system description in terms of *monomials* and *posynomials*, usually obtained by means of data fitting. Although it has been successfully applied in the field of electronic circuits optimization, it has not been applied in this case because it was difficult to model all the several interrelationships between the loss terms.
- Some numerical methods can be applied to function minimization, specially in case of non-linear systems. However, many of them rely the search algorithm on the error and its derivatives. The main disadvantage of most of them is that in case of a function having multiple relative minimums, they may lock on one of them without assuring that it is the absolute minimum.
- *Genetic algorithms* is an special kind of numerical method for function minimization. In this case, several input vectors (W_P, W_N) are randomly scattered throughout the design space and the results are appropriately processed to generate another generation of input vectors. This way, the *genetic algorithms* try to find the target function absolute minimum, in spite of finding some local minimums in the way. Hence, this has been the optimization method used to obtain the optimized designs of the power MOSFETs (and their associated drivers), trying to minimize the power losses.

In the following, a design example of the power MOSFETs of a determined buck converter is presented. The considered buck converter presents the following characteristics: $V_{bat} = 3.6\,V$, $V_o = 1\,V$, $I_o = 100\,mA$, $L = 50\,nH$, $C_o = 30\,nF$, $f_s = 50\,MHz$; which results in DCM operation.

In Fig. 3.63 the evolution of the total power losses (related to the power MOSFETs and their drivers) is shown as a function of power transistors channel width (W_P, W_N). In this case, a coarse sweep of both variables was shown to better show the evolution of the resulting surface. However, all the obtained results by the genetic algorithm (used to find the optimum design) are depicted as scattered

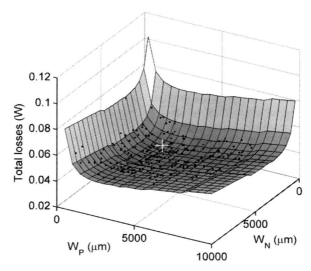

Fig. 3.63 Total power losses related to transistors design, as a function of W_N and W_P. Both results from a complete design space sweep as well as from the genetic algorithm application, are depicted. The optimum design is marked with a *white cross*

black points throughout the design space, rather concentrating around the minimum point. The optimum design is marked by means of a white cross.

It is observed that, in this case, the target function (this is, the power losses) presents a single minimum, thus other optimization procedures rather than genetic algorithms might be used. Nevertheless, in other cases this situation could not be as clear.

The optimized design results in $W_P = 3{,}092\,\mu\text{m}$ and $W_N = 2{,}913\,\mu\text{m}$ power transistors, and their corresponding drivers present the 7.59 and 7.48 as tapering factors, respectively. The overall losses are 37.1 mW.

Figure 3.64 exposes detailed information about the power losses distribution, corresponding to the optimized design.

It is observed that power losses corresponding to PMOS and NMOS are somewhat equalized, as well as the balance between the overall switching losses (22.1 mW) and the conduction losses (15 mW). It is interesting to note that this losses distribution matches the trend shown by the results from Fig. 2.24a, where the major source of switching losses were the drivers.

To summarize, in this section, a method to model the switching losses of the power MOSFETs has been presented. Somehow, two different loss mechanisms that contribute to switching losses are exposed, that account for the loss dependencies on the transistor channel width, the switching inductor current, the switching voltage excursion and the switching interval duration (which is directly related to the driver design and its output current capability). Another important feature of such model is that for a part of the switching losses evaluation, it is not required to know the gate

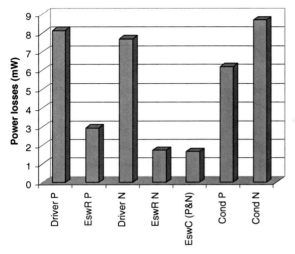

	Power (mW)
Driver P	8,14
E_{swR} P	2,93
Driver N	7,7
E_{swR} N	1,72
E_{swC} (P&N)	1,65
Cond P	6,2
Cond N	8,7

Fig. 3.64 Details of the losses distribution, corresponding the optimized design of the power MOS-FETs and their associated drivers

terminal parasitic resistor, whose value is difficult to predict in case of a standard CMOS implementation, because its dependence on the final layout design.

Additionally, the loss model is suitable for the concurrent design of all the power transistors and their associated drivers. No especial simulations to measure the switching losses of the power MOSFETs (and contrast them against the model results) have been carried out, because it is difficult to separate them from the conduction power losses.

It is not an analytical model (which would be the ideal situation), but it relies on the results of several simple simulations. As a result, an analytical design optimization is precluded and, inevitably, numerical methods have been used to minimize the resulting power losses. In this context, *genetic algorithms* are proposed to optimize the transistor design, as they offer a greater chance to find the global minimum of the target function, and a good trade-off between the computing time consumption and the results accuracy.

Chapter 4
Buck Converter Design Space Exploration with Detailed Component Models

Abstract This chapter covers the review of the Buck converter design space exploration, already presented in Chap. 2. However, in this case, the detailed models presented in Chap. 3 are used to model all the converter components, taking into account their microelectronic implementation by means of a standard CMOS process. The results of such exploration yield and optimized design of the Buck converter in terms of power efficiency and occupied area, and all the details corresponding to each component design are provided. Finally, important conclusions on the feasibility of the microelectronic implementation of a Buck converter are discussed from the results.

The design space exploration has been run using the technical parameters from the 0.25 μm mixed-signal process from UMC. The values of all these parameters are gathered in Table 4.1, classified by the component design in which they are used. As previously explained, most of the data is obtained from simulation sweeps, specially as power MOSFETs and drivers design regards.

In case of the inductor design, the *skin-effect* due to high frequency operation has been included [26]. This results in a reduction of the effective cross section of the bonding wire which, in turn, produces an increase of the equivalent series resistance. Expression (4.2) is used to compute the bonding-wire resistivity (ζ) (by means of the *skin-depth* δ), including the cross-section reduction as a function of the switching frequency (to be more accurate, the spectral composition of the inductor current waveform should be considered; here, just a coarse approach is used by considering a single tone at the switching frequency).

$$\delta = \sqrt{\frac{1}{\pi f_s \sigma \mu_0 \mu_r}} \qquad (4.1)$$

$$\zeta = \frac{1}{\sigma \pi (2r\delta - \delta^2)} \qquad (4.2)$$

Figures 4.1 and 4.2 show the results of the global design space exploration, regarding power efficiency. As in Chap. 2, several frames are presented corresponding to different switching frequency values (f_s), and power efficiency surfaces are

G. Villar Piqué, E. Alarcón, *CMOS Integrated Switching Power Converters*,   121
DOI 10.1007/978-1-4419-8843-0_4, © Springer Science+Business Media, LLC 2011

Table 4.1 Design parameters and technical information used in the Buck converter design space exploration (UMC 0.25 μm mixed-signal process)

Application parameters			
Battery voltage (V_{bat})	3.6 V		
Output voltage (V_o)	1 V		
Output current (I_o)	100 mA		
Maximum output voltage ripple (ΔV_o)	50 mV		
Inductor design			
r	12.5 μm	p	50 μm
σ	41×10^6 S/m	R_{bp}	13 mΩ
$\mu_{r\_conductor}$	0.99996	μ_0	$4\pi \times 10^{-7}$ H/m
μ_{r_media}	1	L tolerance	3%
γ_{LA}	1	γ_{LR}	10
Output capacitor design (core transistors)			
V_{gs}	1 V	C_{ox}	6.27 fF/μm^2
μ_N	376 cm^2/(sV)	V_{term}	25 mV
α	1/12	γ	12
a	0.8 μm	b	0.8 μm
R_{poly}	2.5 Ω	R_{MET1}	53 mΩ
R_{MET2}	53 mΩ	R_{MET3}	53 mΩ
δ_{cont}	1/0.72 cont./μm	δ_{via1}	1/0.76 Via1/μm
δ_{via2}	1/0.76 Via2/μm	R_{cont}	5 Ω/cont.
R_{via1}	3.5 Ω/via1	R_{via2}	3.5 Ω/via2
V_{TN}	0.17 V		
Power MOSFETs and drivers design (IO transistors)			
L_{min}	0.35 μm	W_{min}	0.3 μm
C_{PAD}	1 pF	t_{d_max}	1.5 ns
t_{di}	Table	t_{dei}	Table
t_{fri}	Table	t_{frei}	Table
Q_i	Table	Q_{ei}	Table

depicted as a function of the inductor and output capacitor values. For any shown frame, gray scale has been used to clarify the surface evolution (the higher the power efficiency, the darker the area), and white zones have been used to mark those designs that produce an output voltage ripple (ΔV_o) higher than the maximum allowed by the considered load (50 mV). Additionally, a black dot has been used to mark the design that would produce the higher power efficiency, and black diamonds are used to mark those designs that yield CCM operation. Data on the upper right corner are the design coordinates (L, C_o, f_s) of the maximum power efficiency design, and occupied area, power efficiency and output ripple corresponding to this design configuration, are noted on the upper left corner. Frames corresponding to lower switching frequency values are not shown since they result in even higher output ripple values, thereby not accomplishing the maximum output ripple constraint.

It is observed that the power efficiency is an increasing function of the inductor value, since an inductance increase reduces the RMS value of the inductor current (and, in turn, of the power MOSFETs current). Only when it produces CCM

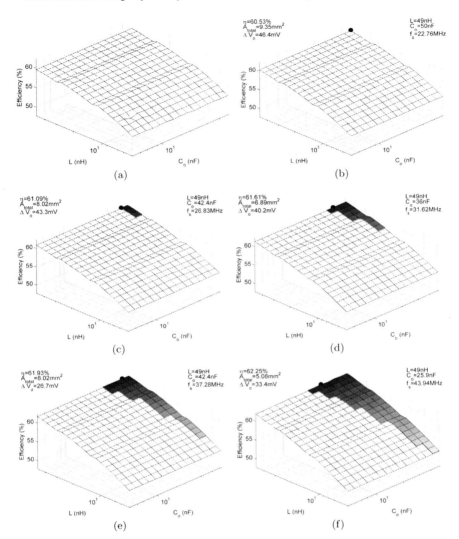

Fig. 4.1 Power efficiency design space exploration results. **a** $f_s = 19.31$ MHz, **b** $f_s = 22.76$ MHz, **c** $f_s = 26.83$ MHz, **d** $f_s = 31.62$ MHz, **e** $f_s = 37.28$ MHz, **f** $f_s = 43.94$ MHz

operation, an inductance increase results in a lower power efficiency (for the same switching frequency, i.e. the same frame) because in that operating mode transistor switching conditions become more lossy. Related to this, it is interesting to note that, when both operating modes exist, the maximum power efficiency is obtained near the border between them, but always in the DCM zone. This is because it represents a very good trade-off between a low current ripple (directly related to its RMS value) and proper switching conditions, due to the DCM operation.

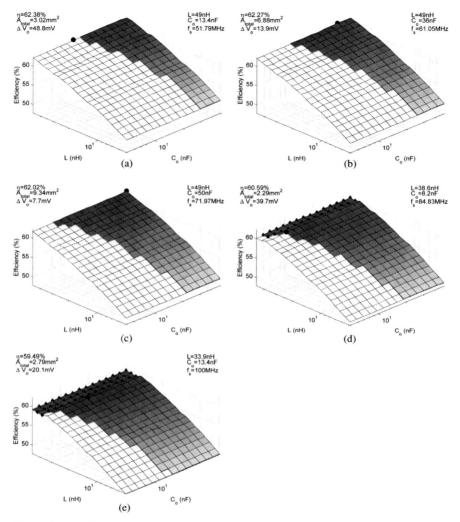

Fig. 4.2 Power efficiency design space exploration results. **a** $f_s = 51.79$ MHz, **b** $f_s = 61.05$ MHz, **c** $f_s = 71.97$ MHz, **d** $f_s = 84.83$ MHz, **e** $f_s = 100$ MHz

As observed in Chap. 2, power efficiency is almost independent of the output capacitor value, and just a variation of its ESR or the particular design of the power MOSFETs and drivers generates an slight change in the power efficiency.

Additionally, power efficiency presents a maximum value as a function of f_s (corresponding to Fig. 4.2a), as the result of the balance of conduction and switching losses, that present opposed trends as a function of the switching frequency.

The occupied area results from the design space exploration are presented in Figs. 4.3 and 4.4 (criteria similar to the power efficiency representation were used).

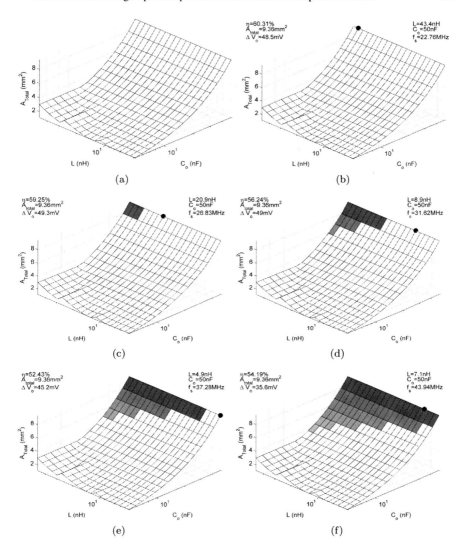

Fig. 4.3 Total occupied area design space exploration results. **a** f_s = 19.31 MHz, **b** f_s = 22.76 MHz, **c** f_s = 26.83 MHz, **d** f_s = 31.62 MHz, **e** f_s = 37.28 MHz, **f** f_s = 43.94 MHz

Regarding area occupancy considerations, it must be taken into account the bonding-wire implementation of the inductor herein proposed, that allows to place the output capacitor as well as the power MOSFETs and their drivers below its floating structure. Thus, the total occupied area is computed as the maximum of the inductor area or the total amount of capacitor, power MOSFETs and drivers area.

$$A_{total} = max[A_L(A_{C_o} + A_{PMOS} + A_{NMOS} + A_{Pdriver} + A_{Ndriver})] \qquad (4.3)$$

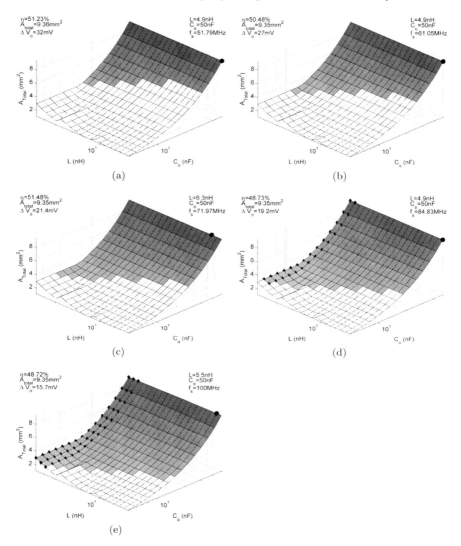

Fig. 4.4 Total occupied area design space exploration results. **a** $f_s = 51.79\,\text{MHz}$, **b** $f_s = 61.05\,\text{MHz}$, **c** $f_s = 71.97\,\text{MHz}$, **d** $f_s = 84.83\,\text{MHz}$, **e** $f_s = 100\,\text{MHz}$

It is observed that the total occupied area becomes almost independent of the switching frequency (since power MOSFETs and drivers area is insignificant in front of the energy-storage components area). It becomes very clear that the switching frequency increase mainly results in a greater expansion of the possible designs area, since the output voltage ripple is reduced.

As a result of expression (4.3), total occupied area is dominated by the output capacitor for high capacitance values, whereas inductor area becomes dominant for low capacitance values. In these cases, inductor area evolution is similar to a sawtooth shape. This is because an increase of the inductance value may be obtained by a larger external side length (which increases the occupied area) or by a higher number of turns. When the later case occurs, usually the external side length is slightly reduced, resulting in a lower area occupancy.

Finally, Figs. 4.5 and 4.6 show the evolution of the defined merit figure that allows to determine the best design trade-off between the power efficiency and the

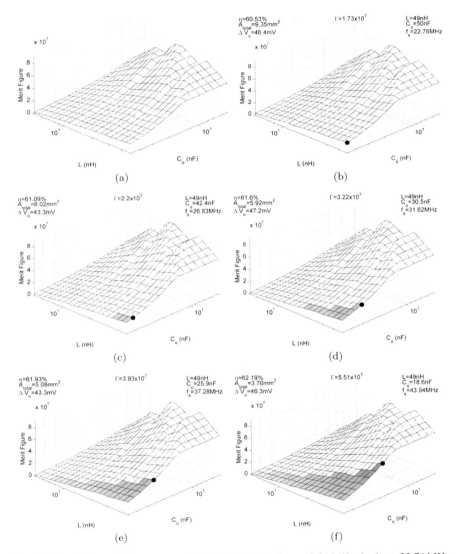

Fig. 4.5 Merit figure design space exploration results. **a** $f_s = 19.31$ MHz, **b** $f_s = 22.76$ MHz, **c** $f_s = 26.83$ MHz, **d** $f_s = 31.62$ MHz, **e** $f_s = 37.28$ MHz, **f** $f_s = 43.94$ MHz

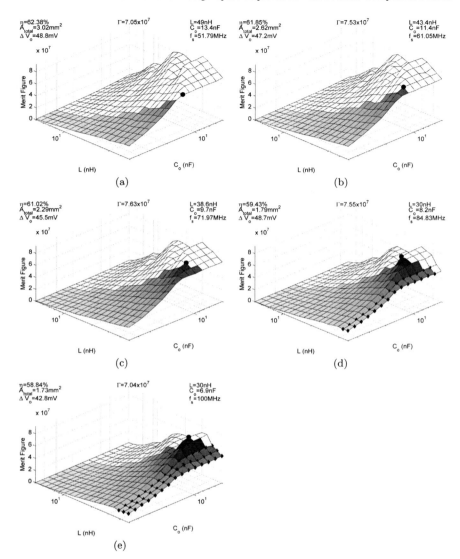

Fig. 4.6 Merit figure design space exploration results. **a** $f_s = 51.79$ MHz, **b** $f_s = 61.05$ MHz, **c** $f_s = 71.97$ MHz, **d** $f_s = 84.83$ MHz, **e** $f_s = 100$ MHz

occupied area, just considering those designs that provide an output ripple lower than the maximum allowed (50 mV). The merit figure definition is presented in (4.4).

$$\Gamma = \frac{(\eta - \eta_{min})^2}{A_{total}} \qquad (4.4)$$

In the merit figure definition, the minimum power efficiency value obtained from the whole design space exploration is substracted from any single power efficiency value obtained for any considered design. This way, the narrow power efficiency span is relatively maximized in front of the total area span (which is much larger). Additionally, to provide a higher relative impact to the power efficiency, this difference is squared. In this case, to simplify the comparison, the resulting Γ value is presented for any frame. Please note that for the sake of a better view, in the merit figure representation, plot axis have been rotated $180°$ respect to the previously presented figures corresponding to power efficiency and occupied area.

The maximum merit figure value corresponds to the frame of Fig. 4.6c. Since power efficiency maximization has been given more priority than occupied area minimization, the maximum merit figure is obtained for a high inductor value. As exposed previously, the converter is DCM operated to minimize power losses. The main effect of the area consideration is that the selected design does not correspond to the maximum power efficiency point, but it is close so as to minimize the total occupied area at the expenses of an slight decrease of the power efficiency (efficiency is decreased by 1.36%, whereas area is reduced by 24.17%).

Additionally, it is observed that for all frames (i.e. all f_s values), the optimum design is found on the edge of the 'forgiven' zone, since the lowest possible values for inductor and capacitor are desired in order to minimize the area occupancy.

Figure 4.7 depicts the evolution of the maximum merit figure value as a function of the switching frequency. This information could be very interesting in the design selection step if further secondary considerations should taken into account, such as the implementation feasibility of the converter control loop. This is, in some cases a very small increase might appear in the maximum merit figure value, for a large

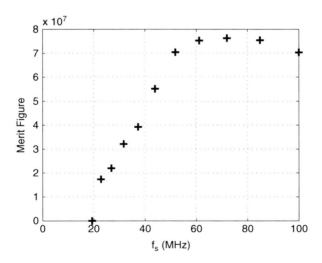

Fig. 4.7 Merit figure evolution as a function of the switching frequency

Table 4.2 Buck converter optimized design main characteristics

Inductor (L)	38.6 nH
Output capacitor (C_o)	9.7 nF
Switching frequency (f_s)	71.97 MHz
On-state duration (T_{on})	3.39 ns
Off-state duration (T_{off})	8.8 ns
Inactivity-state duration (T_i)	1.71 ns
Operating mode	DCM
Maximum inductor current (I_{L_max})	228 mA
Total power losses (P_{losses})	63.9 mW
Power efficiency (η)	61.02%
Total occupied area (A_{total})	2.29 mm^2
Output voltage ripple (ΔV_o)	45.5 mV

increase of the switching frequency, hence resulting in a much more difficult control circuitry implementation (for an small increase of the converter properties).

Tables 4.2, 4.3, 4.4 and 4.5 summarize all the relevant information about the optimized power converter design.

Additionally, Figs. 4.8 and 4.9 show the power losses breakdown and area distribution of the selected design.

As observed, some kind of balance between switching losses and conduction losses is found. Specially interesting is the fact that the power transistors optimization lead to a NMOS switch larger than the PMOS switch, resulting in a large difference between their on-resistance (which may seem unconventional). The reason for that is that the RMS value of the current flowing through the NMOS is much greater than the flowing through the PMOS. Consequently, the overall power losses resulting from any transistor (and their associated drivers) become balanced: excluding the capacitive switching losses (that are consequence of both transistors designs),

Table 4.3 Inductor optimized design characteristics

Inductor design	
Number of turns (n_L)	4
External side length (s_{ext})	2.3 mm
Occupied area (A_L)	2.29 mm^2
ESR (R_L)	1.473 Ω
RMS current (I_L)	123.3 mA
Power losses (P_{L_cond})	22.4 mW

Table 4.4 Output capacitor optimized design characteristics

Output capacitor design	
Number of cells (n)	6,068
Single cells channel length (L_{ch})	2.39 μm
Single cells channel width (W)	106.21 μm
Occupied area (A_{C_o})	2.07 mm^2
ESR (R_{C_o})	19.2 mΩ
RMS current (I_{C_o})	23.3 mA
Power losses (P_{C_o})	10.4 μW

Table 4.5 Buck converter power switches optimized design characteristics

PMOS and P-driver design	
MOSFET channel width (W_{power_MOS})	2, 435.9 μm
On-resistance (R_{on})	1.78 Ω
RMS current (I_{PMOS})	65 mA
Conduction losses (P_{cond})	7.5 mW
Resistive switching losses	4 mW
Driver number of inverters (n)	4
Driver tapering factor (f)	6.13
Minimum inverter PMOS channel width (W_p)	1.43 μm
Driver propagation delay (t_d)	0.99 ns
Driver output fall-rise time (t_{fr})	477 ps
Driver power losses (P_{driver})	8.1 mW
NMOS and N-driver design	
MOSFET channel width (W_{power_MOS})	2, 934.6 μm
On-resistance (R_{on})	0.74 Ω
RMS current (I_{PMOS})	104.8 mA
Conduction losses (P_{cond})	8.1 mW
Resistive switching losses	2.1 mW
Driver number of inverters (n)	4
Driver tapering factor (f)	6.4
Minimum inverter PMOS channel width (W_p)	1.45 μm
Driver propagation delay (t_d)	1.05 ns
Driver output fall-rise time (t_{fr})	501 ps
Driver power losses (P_{driver})	9.6 mW
PMOS and NMOS	
Capacitive switching losses	2.1 mW
Total switching power losses (P_{sw})	25.9 mW
Total conduction power losses	15.6 mW
Total power losses	41.5 mW
Total occupied area	0.008 mm^2

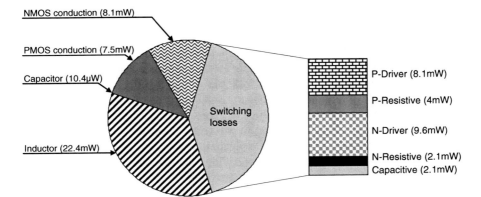

Fig. 4.8 Power losses distribution of the selected Buck converter design

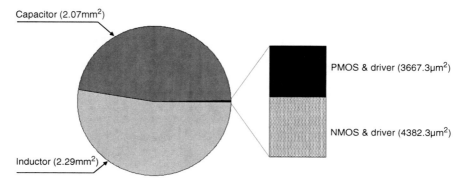

Fig. 4.9 Occupied area distribution of the selected Buck converter design

the total power losses related to the PMOS and NMOS power switches are 19.6 and 19.8 mW, respectively.

Since transistors RMS current values are dependent on the output current and the input-to-output voltage ratio, it must be taken into account that the design is only optimized for the considered application, and it could result in poorer efficiency under other operating conditions. Therefore, in case of attempting to implement a converter to provide different output voltage values, the design space exploration should be carried out for each different output voltage and then, the global efficiency should be evaluated to select the appropriate design.

However, it is still noted that the major source for power losses are the inductor conduction losses. Hence, an improvement on the inductor design would be required to reduce its ESR by reducing the conductor resistivity or using another inductor topology.

As occupied area regards, it is dominated by the inductor area, although some kind of balance between both terms of expression (4.3), which is natural since all not occupied area below the inductor would be lost. In fact, it is foreseen that in case of a more fine design space exploration, inductor area would be the same as the area occupied by the output capacitor plus the power switches and drivers. Again, the energy storage components account for almost the total area.

As a final comment of the results drawn for the refined design space exploration carried out in this chapter, it can be stated that although being optimized through the design space exploration, the obtained results appear somewhat disappointing because of the obtained poor power efficiency for a circuit that, theoretically, should provide a 100% power efficiency, despite still being better than that provided by a linear voltage converter (for the same application parameters). Moreover, the maximum obtained power efficiency (corresponding to the design of Fig. 4.2a), is still considered too low for an efficient voltage converter.

Inevitably, as exposed in the design procedure proposed in Sect. 2.1, because of this unsatisfactory performance factors results, an alternative converter topology should be considered to implement the desired functionality with improved performance as regards power efficiency.

Chapter 5
3-Level Buck Converter Analysis and Specific Components Models

Abstract From the results of the Buck converter design space exploration, the implementation of a different converter topology is proposed in this chapter. Having discussed some of the required characteristics to reduce the power losses, and consequently increase the converter power efficiency, the 3-level Buck converter presented by Meynard and Foch in 1992 (*Power Electronics Specialists conference, 1992. PESC'92 Record, 23rd Annual IEEE*, Toledo), is selected. Then, the analysis of the converter operation is presented since it is required for its microelectronic implementation and the low power operation. To develope that, the converter operation analysis is subdivided as a function of the output voltage ($V_o \lesssim V_{bat}/2$), and the operating mode (CCM or DCM). From this analysis, the main characteristics are compared with those from the classical Buck converter. After that, a self-driving scheme to supply the power drivers and make possible the use of shorter channel transistors (core transistors), is proposed. Finally, detailed additional component models required to carry out the corresponding design space exploration are provided, as well as some design considerations.

5.1 Motivation

The analysis of the design space exploration results obtained for the Buck converter, specially in terms of power efficiency, lead to the pursuit of an improved converter topology able to yield better efficiency results, for the same target application.

In order to find it, the most important loss mechanisms must be identified to attempt to reduce them down. From the results of Fig. 4.8, it is observed that conduction losses (specially from the inductor series resistance) become a very important contributor to power losses. These are due to the inductor and power switches physical design (which is supposed to be optimized) and materials (which are imposed by the chosen technological process).

In this scenario, the only way to reduce the conduction losses would be to reduce the RMS value of the inductor current I_L (which, in turn, is a significative indicator of the RMS value of the current flowing through each of the power switches, too). Reminding expressions (2.17) from Chap. 2, it is observed that it depends

G. Villar Piqué, E. Alarcón, *CMOS Integrated Switching Power Converters*,   133
DOI 10.1007/978-1-4419-8843-0_5, © Springer Science+Business Media, LLC 2011

on the application parameters (I_o, V_o, V_{bat}, that are inevitably determined by the considered application), and two of the design variables (f_s and L, which are supposed to be properly selected by means of the design space exploration). In conclusion, it would seem that nothing could be done to reduce the I_L value.

Nevertheless, the V_{bat} value that appears on those expressions identifies the maximum value of the chopped signal at the x-node, that in case of a classical Buck converter coincides with the input (battery) voltage V_{bat}. Hence, in a strict way, the inductor RMS current could be expressed by:

$$I_L^2 = \sqrt{\frac{8I_o^3 V_o(V_{x\_max} - V_o)}{9V_{x\_max}Lf_s}} \longrightarrow DCM \qquad (5.1)$$

$$I_L^2 = \frac{V_o^2(V_{x\_max} - V_o)^2 V_{x\_max}}{12L^2 f_s^2 V_{x\_max}^3} + I_o^2 \longrightarrow CCM \qquad (5.2)$$

where V_{x_max} is the upper level of the v_x voltage, which it is not required to coincide with the input voltage (V_{bat}). In fact, for the step-down operation to work, it is just needed that $V_{x\_max} > V_o$. In Fig. 5.1, the RMS value of the inductor current (I_L) as a function of the V_{x_max} voltage is depicted. As observed, the minimum limit is the output voltage because some positive difference between V_{x_max} and V_o is required for the inductor current to flow towards the converter output.

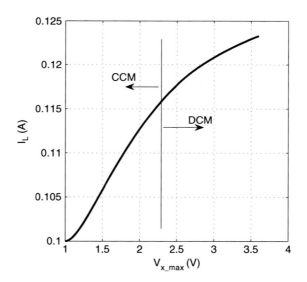

Fig. 5.1 Inductor current RMS evolution as a function of the upper level of the v_x voltage, for a classical Buck converter. Used parameters: $V_o = 1$ V, $I_o = 100$ mA, $f_s = 72$ MHz and $L = 38.6$ nH

The minimum I_L value coincides with the output current (in the example, $I_o = 100\,mA$), which implies that the inductor almost does not present any ripple despite the switching cycle duration, since the voltage difference across the inductor is almost zero. Thus, I_L is an increasing function of V_{x_max}.

Therefore, a possible way to reduce the conduction losses could be a converter topology that reduces the V_{x_max} value close to V_o. This reflexion suggests the use of some kind of multilevel step-down converter.

From all the possible multilevel converter topologies, the one developed by Meynard and Foch in 1992 [111] has been selected, whose schematic is depicted in Fig. 5.2.

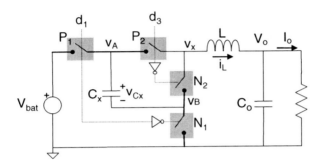

Fig. 5.2 Ideal scheme of the 3-level Buck converter

The Meynard switching cell was firstly developed for High-Voltage Power Conversion applications (specially voltage inverters), where the large voltage drop that must be blocked by the power switches during their off-state is a serious reliability issue. Hence, this cell presents the ability to reduce the voltage across the semiconductors in such situations, by means of the use of a capacitor which provides half of the input voltage, due to natural energy balance. Additionally, it yields a reduction of the harmonic spectrum of the circuit signals that it is interesting for some applications. The circuit operation is based on applying 2 different switching signals (d_1 and d_3, in the scheme) with the same duty cycle, whose phases have been shifted half the switching period.

As it will be shown afterwards, such operation results in the possibility to get 3 different levels at the switching x-node: V_{bat}, 0 and $V_{bat}/2$, in a first order approach. Because of this, in this work, it will be called as the *3-level Buck converter* from now on, to distinguish it from the *classical Buck converter*. Further details on this converter operation can be found in the literature.

However, this topology has been more rarely used as a low-voltage regulator, and even more rare is its use in low power applications, where the DCM operation could not only appear, but become the main operating mode. As a result, in this chapter, the DCM operation will be analyzed in detail, as well as the CCM operation. Furthermore, the possibility to use a low C_x value (as imposed in an IC implementation), and the consequent pseudo-resonant behavior, will also be included in the analysis.

Other benefits that are really appreciated when focusing the integration of switching power converters are listed in the following:

- As it will be demonstrated, lower output voltage ripple results from this converter operation, for the same values of switching frequency, inductance and output capacitance. Thus, a greater area of possible designs is expected in the design space exploration.
- The presence of lower voltage levels throughout the circuit allows the use of shorter-channel transistors (that usually present a lower breakdown voltage), which reduces their on-resistance as well as their parasitic capacitances (related to switching losses).
- Because of the narrower voltage excursions, lower capacitive switching losses could result.
- The lower difference between V_{x_max} and V_o results in lower slopes from the inductor current waveform. This could produce a relatively lower I_{L_max} value, that generates lower resistive switching losses.
- Both previous characteristics also reduce the power switches electrical stress, thereby increasing the system reliability.

However, some disadvantages should also be taken into account:

- First of all, the presence of an additional capacitor (C_x, in Fig. 5.2) increases the occupied area, as well as the conduction losses (due to its series resistance).
- Along any complete switching cycle, 4 power transistors (with their corresponding drivers) switch their state. Thus, an increase of switching losses could result.
- In any of the converter states, 2 power transistors are series connected, which would increase conduction losses due to the higher total on-resistance.
- The required switching and control schemes become more complex due to the four power switches to be controlled.

After considering all the previous advantages and disadvantages, a clear conclusion on the benefits of the use of the Meynard 3-level converter can not be found. As a result of this, in the following sections a detailed analysis and comparison with the Buck converter is exposed to obtain a suitable model that fits the design space exploration structure, to evaluate all the required performance factors, and eventually obtain the optimized design.

5.2 Ideal Analysis

First of all, all four different states of the circuit will be presented as well as the corresponding switching sequence. Then, a detailed analysis will be carried out to obtain significant waveforms and characteristics such as output ripple and inductor current RMS value. From the analysis results, a comparison between the 3-level and the classic Buck converter will be presented.

5.2.1 Basic Operation

The following analysis is based on the ideal circuit scheme shown in Fig. 5.2. From that circuit, four different basic configurations are derived from the combined value of d_1 and d_3 switching signals. If switches are considered closed when their corresponding switching signals are high, the main configurations and associated signal combinations are obtained, as summarized in Fig. 5.3.

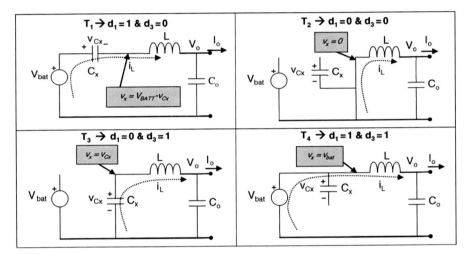

Fig. 5.3 4 main states of the 3-level Buck converter

Additionally, in order to clarify the circuit configurations, inductor current has been depicted as well as its sense, which is supposed not to reverse in any case; and the resulting voltage value at the x-node is marked in gray boxes.

Since both switching signals (d_1, d_3) present the same duty cycle ($D_1 = D_3$) and are applied with a 180° phase, in the following, the resulting switching sequences for duty cycles lower and higher than 50% are described.

In case of a duty cycle value lower than 50% the resulting switching sequence is $T_1 \rightarrow T_2 \rightarrow T_3 \rightarrow T_2$, as it is observed in Fig 5.4.

When duty cycle is higher than 50%, the circuit functionality becomes as depicted in Fig. 5.5, being the switching sequence $T_4 \rightarrow T_1 \rightarrow T_4 \rightarrow T_3$.

As a first approach, v_x voltage is considered to remain constant along T_1 and T_3 states, provided that C_x capacitor is large enough. Additionally, v_x at these states ($V_{bat} - v_{C_x}$ and v_{C_x}, respectively) is expected to be $V_{bat}/2$ because of the circuit natural balance through T_1 and T_3 states.

Fig. 5.4 3-level converter ideal waveforms, in case of switching signals duty cycle lower than 50%

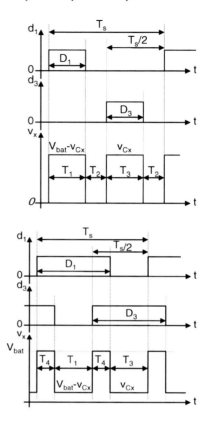

Fig. 5.5 3-level converter ideal waveforms, in case of switching signals duty cycle higher than 50%

As a function of the C_x capacitor, the inductor value and the output current, two different operating modes are possible for the 3-level Buck converter: continuous conduction mode (CCM), and discontinuous conduction mode (DCM). Hence, since two different switching sequences are possible as a function of the switching signals duty cycle, four different situations should be analyzed.

A significant improvement over the classical model of the 3-level Buck converter is the inclusion of the effect of a small C_x capacitor (in order to make more feasible its microelectronic implementation) which results in some resonant behavior between this capacitor and the inductor that should be taken into account in the evaluation of significant circuit parameters, such as the output voltage ripple.

In this case, parasitic series resistances from any of the components have not been considered to avoid complexity in the analytical treatment of the resulting expressions.

Provided that self-balancing of the C_x capacitor voltage exists from T_1 and T_3 states, only expressions from the first semi-period of the switching sequence are necessary to get the desired functional parameters, since v_x voltage and i_L current will present the same waveforms in both semi-periods (the only difference is that during T_1, C_x capacitor is charged while it is discharged along T_3).

5.2.2 *3-Level Buck Converter Analytical Expressions*

5.2.2.1 DCM Operation and $D_1 < 0.5$

During the T_1 state, the inductor and the C_x capacitor are series-connected, resulting in an increase of the inductor current (i_L) from 0, because of the DCM operation.

In this case, the voltage across the inductor is $V_{bat} - v_{C_x} - V_o$, being V_0 the initial value of v_{C_x}. Thus, the i_L expression corresponding to the T_1 state is:

$$i_L(t) = (V_{bat} - V_0 - V_o)\sqrt{\frac{C_x}{L}} \sin\left(\frac{t}{\sqrt{LC_x}}\right) \tag{5.3}$$

And the charge stored in the C_x capacitor at the end of the T_1 state is:

$$Q_1 = (V_{bat} - V_0 - V_o)C_x \left[1 - \cos\left(\frac{T_1}{\sqrt{LC_x}}\right)\right] \tag{5.4}$$

To keep the energy balance through the whole switching period, the total energy supplied by the battery must be equal to the total energy supplied to the output load, since no energy losses are considered.

$$V_{bat}Q_1 = V_o I_o T_s \tag{5.5}$$

$$V_{bat}(V_{bat} - V_0 - V_o)C_x \left[1 - \cos\left(\frac{T_1}{\sqrt{LC_x}}\right)\right] = \frac{V_o I_o}{f_s} \tag{5.6}$$

And from the previous expression, the necessary T_1 state duration can be found in order to keep the total energy balance:

$$T_1 = \sqrt{LC_x} \arccos\left(1 - \frac{V_o I_o}{f_s(V_{bat} - V_o - V_0)V_{bat}C_x}\right) \tag{5.7}$$

The initial condition for the C_x capacitor voltage (V_0) can be obtained if the v_{C_x} voltage is supposed to vary around the $V_{bat}/2$ value. The total v_{C_x} voltage change along the T_1 state (Δv_{C_x}) can be found from the total stored charge (Q_1).

$$V_0 = \frac{V_{bat}}{2} - \frac{\Delta v_{C_x}}{2} = \frac{V_{bat}}{2} - \frac{Q_1}{2C_x} \tag{5.8}$$

In expression (5.8), a minus sign is used since, from the expected inductor current direction in T_1, the initial capacitor charge is supposed to be less than at the end of T_1 (it is considered that C_x capacitor is charged along T_1). And from previous expression (5.5) a simple expression for V_0 can be obtained.

$$V_0 = \frac{V_{bat}}{2} - \frac{V_o I_o}{2 C_x f_s V_{bat}} \qquad (5.9)$$

Additionally, the expression of the inductor current at the end of T_1 is given because it is the initial condition of the inductor current in the T_2 state analysis.

$$I_{T_1} = (V_{bat} - V_o - V_0)\sqrt{\frac{C_x}{L}} \sin\left(\frac{T_1}{\sqrt{LC_x}}\right) \qquad (5.10)$$

During T_2, x-node is directly connected to ground ($v_x = 0$) producing a voltage drop across the inductor of $-V_o$. Thus, inductor current will decrease until 0, as expressed in (5.11).

$$i_L(t) = I_{T_1} - \frac{V_o}{L}t \qquad (5.11)$$

And the T_2 phase duration is:

$$T_2 = \frac{I_{T_1} L}{V_o} \qquad (5.12)$$

As regards the output voltage ripple, two different cases are possible, depending on whether the I_{T_1} value is higher or not than the output current, as exposed in Fig. 5.6.

In the first case ($I_{T_1} > I_o$), the output voltage ripple is obtained from the following expressions:

$$T'_{1m} = \sqrt{LC_x} \arcsin\left(\frac{I_o}{V_{bat} - V_o - V_0}\sqrt{\frac{L}{C_x}}\right) \qquad (5.13)$$

$$T'_2 = \frac{(I_{T_1} - I_o)L}{V_o} \qquad (5.14)$$

$$Q_B = I_{T_1} T'_2 - \frac{V_o}{2L}T_2'^2 \qquad (5.15)$$

$$Q_A = (V_{bat} - V_o - V_0)C_x \left(\cos\left(\frac{T'_{1m}}{\sqrt{LC_x}}\right) - \cos\left(\frac{T_1}{\sqrt{LC_x}}\right)\right) \qquad (5.16)$$

$$\Delta V_o = \frac{Q_A + Q_B - I_o \left(T'_2 + T_1 - T'_{1m}\right)}{C_o} \qquad (5.17)$$

For the second case, the following expressions are used to evaluate the output voltage ripple:

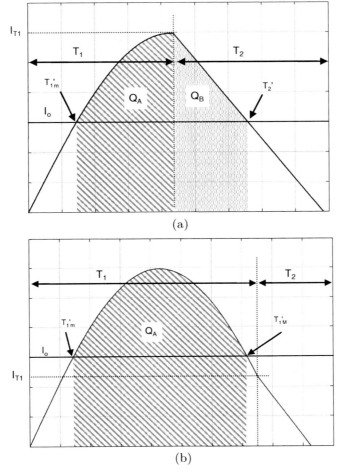

Fig. 5.6 Output capacitor charge evaluation depending on the relationship between I_{T_1} and I_o:
a $I_{T_1} > I_o$ and **b** $I_{T_1} < I_o$

$$T'_{1m} = \sqrt{LC_x} \arcsin \left(\frac{I_o}{V_{bat} - V_o - V_0} \sqrt{\frac{L}{C_x}} \right) \tag{5.18}$$

$$T'_{1M} = \sqrt{LC_x}\pi - T'_{1m} \tag{5.19}$$

$$Q_A = (V_{bat} - V_o - V_0)C_x \left(\cos\left(\frac{T'_{1m}}{\sqrt{LC_x}}\right) - \cos\left(\frac{T'_{1M}}{\sqrt{LC_x}}\right) \right) \tag{5.20}$$

$$\Delta V_o = \frac{Q_A - I_o\left(T'_{1M} - T'_{1m}\right)}{C_o} \tag{5.21}$$

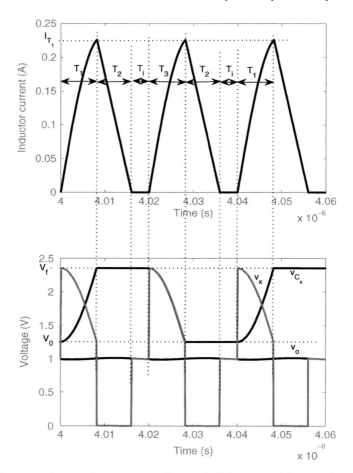

Fig. 5.7 Representative waveforms corresponding to a DCM operated 3-level Buck converter, in case of a duty cycle lower than 50%

In Fig. 5.7 the most significant waveforms corresponding to a DCM operated 3-level Buck converter are depicted for the case of $D_1 < 0.5$: inductor current (i_L), output voltage (v_o), C_x capacitor voltage (v_{C_x}) and x-node voltage (v_x). The converter parameters used in this MATLAB simulation are: $V_{bat} = 3.6$ V, $V_o = 1$ V, $I_o = 100$ mA, $L = 35$ nH, $C_o = 30$ nF, $C_x = 1$ nF and $f_s = 25$ MHz.

In the figure, T_i is used to identify those inactivity states for which no current flows through the inductor.

Due to the low C_x value, it is observed some resonant behavior in the inductor current, as well as important voltage swing on v_{C_x} around the $V_{bat}/2 = 1.8$ V value.

In order to evaluate the RMS value of the current that flows through each of the components (required to compute their corresponding conduction losses), the inductor current waveform is split into different parts corresponding to each of the

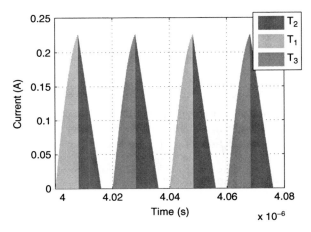

Fig. 5.8 Inductor current waveform division to evaluate the RMS value of the current flowing through the different circuit components

different converter states. This division can be observed in Fig. 5.8. Then, the RMS value of those sub-waveforms is calculated assuming they are the only current waveforms along the whole switching cycle.

Therefore, the RMS value of the inductor current corresponding to the T_1 state is first computed.

$$I_{LT_1RMS} = \sqrt{f_s \left[\frac{C_x}{2L} (V_{bat} - V_o - V_0)^2 \left(T_1 - \sqrt{LC_x} \sin\left(\frac{T_1}{\sqrt{LC_x}}\right) \cos\left(\frac{T_1}{\sqrt{LC_x}}\right) \right) \right]} \quad (5.22)$$

Since the current waveform along T_3 is exactly the same as during T_1, $I_{LT_3RMS} = I_{LT_1RMS}$. And the current RMS value along T_2 is:

$$I_{LT_2RMS} = \sqrt{f_s \left[I_{T_1}^2 T_2 + \frac{V_o^2 T_2^3}{3L^2} - \frac{I_{T_1} V_o T_2^2}{L} \right]} \quad (5.23)$$

Then, the RMS value of the current through each component is straightforwardly evaluated.

$$I_L^2 = 2 \left(I_{LT_1RMS}^2 + I_{LT_2RMS}^2 \right) \quad (5.24)$$

$$I_{P_1}^2 = I_{LT_1RMS}^2 \quad (5.25)$$

$$I_{P_2}^2 = I_{LT_1RMS}^2 \quad (5.26)$$

$$I_{N_1}^2 = 2I_{LT_2RMS}^2 + I_{LT_1RMS}^2 \tag{5.27}$$

$$I_{N_2}^2 = 2I_{LT_2RMS}^2 + I_{LT_1RMS}^2 \tag{5.28}$$

$$I_{C_x}^2 = 2I_{LT_1RMS}^2 \tag{5.29}$$

As it will be justified later in this chapter, this computation method (by means of the inductor current waveform subdivision) becomes very useful when evaluating the power MOSFETs conduction losses for the 3-level Buck converter.

5.2.2.2 CCM Operation and $D_1 < 0.5$

In this case, during T_1 state the inductor current rises from I_0 while it charges the C_x capacitor, and the voltage across the inductor is initially $V_{bat} - V_o - V_0$. The expression corresponding to the evolution of i_L along T_1 state is as follows:

$$i_L(t) = (V_{bat} - V_o - V_0)\sqrt{\frac{C_x}{L}} \sin\left(\frac{t}{\sqrt{LC_x}}\right) + I_0 \cos\left(\frac{t}{\sqrt{LC_x}}\right) \tag{5.30}$$

And the corresponding inductor current value at the end of T_1 is:

$$I_{T_1} = (V_{bat} - V_o - V_0)\sqrt{\frac{C_x}{L}} \sin\left(\frac{T_1}{\sqrt{LC_x}}\right) + I_0 \cos\left(\frac{T_1}{\sqrt{LC_x}}\right) \tag{5.31}$$

The inductor current expression during T_2 state is:

$$i_L(t) = I_{T_1} - \frac{V_o}{L}t \tag{5.32}$$

Its integration along the T_2 state duration provides the charge stored in C_x during this state:

$$Q_2 = \int_0^{T2} \left(I_{T_1} - \frac{V_o}{L}t\right)dt = I_{T_1}T_2 - \frac{V_o}{2L}T_2^2 \tag{5.33}$$

To obtain the values of T_1, I_0 and V_0, 3 different equations are to be solved (by means of numerical methods, since no analytical solutions could be found).

1. The first condition to be accomplished is that the current increments along T_1 and T_2 should be the same:

$$0 = (V_{bat} - V_o - V_0)\sqrt{\frac{C_x}{L}} \sin\left(\frac{T_1}{\sqrt{LC_x}}\right) + I_0\left[\cos\left(\frac{T_1}{\sqrt{T_1}}\right) - 1\right] - \frac{V_o}{L}\left(\frac{1}{2f_s} - T_1\right) \tag{5.34}$$

where the equivalence $T_2 = \frac{1}{2f_s} - T_1$ was used.

2. The second condition is that the average value of the inductor current must be equal to the output current I_o.

$$I_o = \frac{Q_1 + Q_2}{T_1 + T_2} \longrightarrow 0 = 2f_s(Q_1 + Q_2) - I_o \tag{5.35}$$

$$0 = 2f_s \left[(V_{bat} - V_o - V_0)C_x \left[1 - \cos\left(\frac{T_1}{\sqrt{LC_x}} \right) \right] + \right.$$
$$\left. +I_0\sqrt{LC_x} \sin\left(\frac{T_1}{\sqrt{LC_x}} \right) + I_{T_1}\left(\frac{1}{2f_s} - 1 \right) - \frac{V_o}{2L}\left(\frac{1}{2f_s} - T_1 \right)^2 \right] \tag{5.36}$$

3. And the third condition is that the initial C_x capacitor voltage (V_0) must be equal to $\frac{V_{bat}}{2} - \frac{\Delta v_{C_x}}{2}$.

$$0 = \frac{V_o + (V_{bat} - V_o)\cos\left(\frac{T_1}{\sqrt{LC_x}} \right) - I_0\sqrt{\frac{L}{C_x}}\sin\left(\frac{T_1}{\sqrt{LC_x}} \right)}{1 + \cos\left(\frac{T_1}{\sqrt{LC_x}} \right)} - V_0 \tag{5.37}$$

As regards the output voltage ripple, its calculation is carried out with the same expressions as in (5.13), (5.14), (5.15), (5.16), (5.17), (5.18), (5.19), (5.20), (5.21), but in the CCM operation case the T'_{1m} value is obtained from the resolution of (5.38)

$$0 = (V_{bat} - V_o - V_0)\sqrt{\frac{C_x}{L}}\sin\left(\frac{T'_{1m}}{\sqrt{LC_x}} \right) + I_0\cos\left(\frac{T'_{1m}}{\sqrt{LC_x}} \right) - I_o \tag{5.38}$$

Figure 5.9 shows the most significant waveforms corresponding to a CCM operated 3-level Buck converter for the case of $D_1 < 0.5$: inductor current (i_L), output voltage (v_o), C_x capacitor voltage (v_{C_x}) and x-node voltage (v_x). The converter parameters used in this MATLAB simulation: $V_{bat} = 3.6\,\text{V}$, $V_o = 1\,\text{V}$, $I_o = 200\,\text{mA}$, $L = 35\,\text{nH}$, $C_o = 30\,\text{nF}$, $C_x = 3\,\text{nF}$ and $f_s = 25\,\text{MHz}$.

To compute the RMS value of the current through each of the converter components, a method similar to the DCM operation counterpart has been used.

The RMS value of $i_L(t)$ due to the T_1 state (which coincides with the one resulting from T_3 state) is expressed as follows:

$$I_{LT_1 RMS} = \left[f_s \left[\frac{T_1}{2}\left[(V_{bat} - V_o - V_0)^2\frac{C_x}{L} + I_0^2 \right] + \right. \right.$$
$$+\frac{\sqrt{LC_x}}{2}\left[I_0^2 - (V_{bat} - V_o - V_0)^2\frac{C_x}{L} \right]\sin\left(\frac{T_1}{\sqrt{LC_x}} \right)\cos\left(\frac{T_1}{\sqrt{LC_x}} \right) +$$
$$\left. \left. +I_0 C_x(V_{bat} - V_o - V_0)\left(1 - \cos^2\left(\frac{T_1}{\sqrt{LC_x}} \right) \right) \right] \right]^{1/2} \tag{5.39}$$

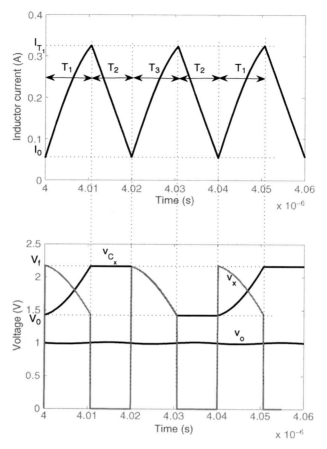

Fig. 5.9 Representative waveforms corresponding to a CCM operated 3-level Buck converter, in case of a duty cycle lower than 50%

And the expression corresponding to the T_2 state is:

$$I_{LT_2 RMS} = \sqrt{f_s \left[I_{T_1}^2 T_2 + \frac{V_o^2 T_2^3}{3L^2} - \frac{I_{T_1} V_o T_2^2}{L} \right]} \qquad (5.40)$$

Then, the RMS current value for each of the components is directly computed, and, as expected, coincides with the expressions found for the DCM case.

$$I_L^2 = 2 \left(I_{LT_1 RMS}^2 + I_{LT_2 RMS}^2 \right) \qquad (5.41)$$

$$I_{P_1}^2 = I_{LT_1 RMS}^2 \qquad (5.42)$$

$$I_{P_2}^2 = I_{LT_1 RMS}^2 \tag{5.43}$$

$$I_{N_1}^2 = 2I_{LT_2 RMS}^2 + I_{LT_1 RMS}^2 \tag{5.44}$$

$$I_{N_2}^2 = 2I_{LT_2 RMS}^2 + I_{LT_1 RMS}^2 \tag{5.45}$$

$$I_{C_x}^2 = 2I_{LT_1 RMS}^2 \tag{5.46}$$

5.2.2.3 DCM Operation and $D_1 > 0.5$

When the output voltage is higher than $V_{bat}/2$ ($D_1 > 0.5$), the first state considered in a switching cycle is T_4, which is followed by T_1, in the first half of the period. In the second semiperiod, T_4 is followed by T_3, resulting in the same voltage and current waveforms with the exception that C_x capacitor is charged during T_1 and discharged during T_3. Therefore, in the following analysis, only the $T_4 \rightarrow T_1$ switching sequence is taken into account and is considered to be repeated, with the aforementioned exception.

In DCM operation the inductor current expression along T_4 is:

$$i_L(t) = \frac{V_{bat} - V_o}{L} t \tag{5.47}$$

Which results in a charge supplied by the input voltage source of:

$$Q_4 = \frac{V_{bat} - V_o}{L} T_4^2 \tag{5.48}$$

And the inductor current value at the end of T_4:

$$I_{T_4} = \frac{V_{bat} - V_o}{L} T_4 \tag{5.49}$$

The inductor current during T_1 state is expressed by (5.50)

$$i_L(t) = (V_{bat} - V_o - V_0)\sqrt{\frac{C_x}{L}} \sin\left(\frac{t}{\sqrt{LC_x}}\right) + I_{T_4} \cos\left(\frac{t}{\sqrt{LC_x}}\right) \tag{5.50}$$

In DCM operation, the duration of T_1 state is determined by the time for which the inductor current reaches 0, starting from I_{T_4}. Hence, the T_1 value is obtained from expression (5.50).

$$T_1 = \sqrt{LC_x} \arctan\left(\frac{-I_{T_4}}{V_{bat} - V_o - V_0}\sqrt{\frac{L}{C_x}}\right) \tag{5.51}$$

And the charge stored in the C_x capacitor throughout T_1 is:

$$Q_1 = (V_{bat} - V_o - V_0)C_x \left[1 - \cos\left(\frac{T_1}{\sqrt{LC_x}} \right) \right] + I_{T_4}\sqrt{LC_x} \sin\left(\frac{T_1}{\sqrt{LC_x}} \right) \quad (5.52)$$

In order to obtain the values of T_4 and V_0, 2 non-linear equations must be solved (by means of numerical methods), which arise from 2 different conditions that must be accomplished.

1. The first condition is that the input energy to the converter along a switching period should be equal the output energy in the same time, provided that no energy losses are considered in these analysis.

$$V_{bat}(2Q_4 + Q_1) = V_o I_o T_s \quad (5.53)$$

$$0 = \frac{V_o I_o}{V_{bat} f_s} - Q_1 - \frac{V_{bat} - V_o}{L} T_4^2 \quad (5.54)$$

2. The second condition is that the v_{C_x} voltage balance along a complete switching period must be zero.

$$V_0 = \frac{V_o + (V_{bat} - V_o)\cos\left(\frac{T_1}{\sqrt{LC_x}} \right) - I_{T_4}\sqrt{\frac{L}{C_x}} \sin\left(\frac{T_1}{\sqrt{LC_x}} \right)}{1 + \cos\left(\frac{T_1}{\sqrt{LC_x}} \right)} \quad (5.55)$$

From the chronogram of Fig. 5.5, the relationship between the D_1 duty cycle and the T_4 state duration is found in expression (5.56).

$$D_1 = T_4 f_s + \frac{1}{2} \quad (5.56)$$

As for $D_1 < 0.5$, two different situations appear regarding the output voltage ripple calculation, as a function of the relationship between the output current (I_o), and the inductor current at the end of T_4 (I_{T_4}). In Fig. 5.10, both situations are depicted.

To compute the output voltage ripple in case of $I_{T_4} > I_o$, expressions from (5.57), (5.58), (5.59), and (5.60) are required.

$$T_4' = \frac{I_o L}{V_{bat} - V_o} \quad (5.57)$$

$$Q_A = \frac{V_{bat} - V_o}{2L} \left(T_4^2 - T_4'^2 \right) \quad (5.58)$$

$$Q_B = (V_{bat} - V_o - V_0)C_x \left[1 - \cos\left(\frac{T_{1M}'}{\sqrt{LC_x}} \right) \right] + I_{T_4}\sqrt{LC_x} \sin\left(\frac{T_{1M}'}{\sqrt{LC_x}} \right) \quad (5.59)$$

$$\Delta V_o = \frac{Q_A + Q_B - I_o \left(T_{1M}' + T_4 - T_4' \right)}{C_o} \quad (5.60)$$

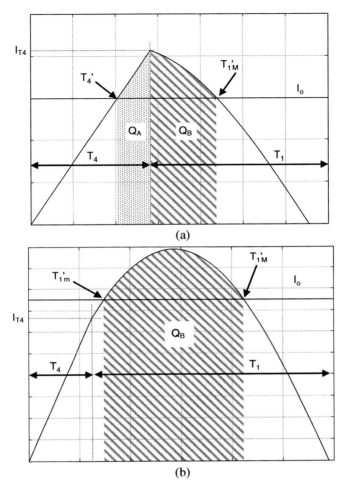

Fig. 5.10 Output capacitor charge evaluation depending on the relationship between I_{T_4} and I_o:
a $I_{T_4} > I_o$ and **b** $I_{T_4} < I_o$

If $I_{T_4} < I_o$, output ripple is computed as follows:

$$Q_B = (V_{bat} - V_o - V_0)C_x \left[\cos\left(\frac{T'_{1m}}{\sqrt{LC_x}}\right) - \cos\left(\frac{T'_{1M}}{\sqrt{LC_x}}\right) \right] +$$

$$+ I_{T_4}\sqrt{LC_x} \left[\sin\left(\frac{T'_{1M}}{\sqrt{LC_x}}\right) - \sin\left(\frac{T'_{1m}}{\sqrt{LC_x}}\right) \right] \tag{5.61}$$

$$\Delta V_o = \frac{Q_B - I_o \left(T'_{1M} - T'_{1m}\right)}{C_o} \tag{5.62}$$

The T'_{1m} and T'_{1M} values are obtained from the expression (5.63). Nevertheless, only the feasible solutions should be used for each case.

$$0 = (V_{bat} - V_o - V_0)\sqrt{\frac{C_x}{L}} \sin\left(\frac{T'_1}{\sqrt{LC_x}}\right) + I_{T_4}\cos\left(\frac{T'_1}{\sqrt{LC_x}}\right) - I_o \qquad (5.63)$$

The waveforms corresponding to the inductor current, x-node voltage, C_x capacitor voltage and output voltage of a DCM operated 3-level Buck converter when its output is higher than half of its input voltage can be found in Fig. 5.11.

The RMS value of the current due to the T_1 state (which coincides with the one resulting from T_3 state), and due to T_4 are obtained.

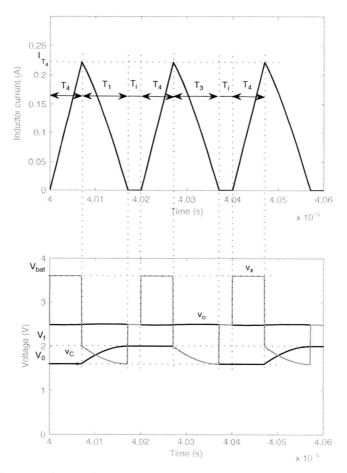

Fig. 5.11 Representative waveforms corresponding to a DCM operated 3-level Buck converter, in case of a duty cycle higher than 50%

5.2.3 3-Level and Classical Buck Converters Comparison

In this section, the comparison between the classical and the 3-level Buck converters is presented in terms of the most significant performance indexes, when focusing on its monolithic integration. The specifications for both converters can be observed in Table 5.1 (since just the ideal functionality is observed, neither transistors and drivers designs were included in the comparison, nor other components non-idealities). In most of the design parameters the same values have been used, targeting an homogeneous comparison. However, it should be taken into account that the 3-level converter requires a higher number of power transistors and drivers, as well as an additional capacitor.

Table 5.1 Compared converters main parameters

Parameter	Value
L	35 nH
C_o	30 nF
f_s	25 MHz
V_{bat}	3.6 V
V_o	1 V
I_o	100 mA

In Figs. 5.13 and 5.14 the operating mode and the duty cycle of the corresponding switching signal are depicted as a function of the output voltage and the C_x capacitor value for the 3-level Buck converter. The transfer function is shown through the d_1 signal duty-cycle (in %).

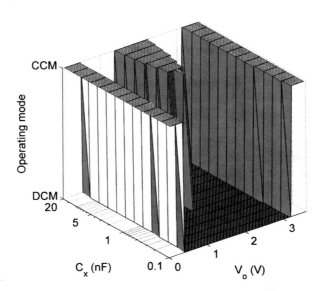

Fig. 5.13 Operating mode of the 3-level Buck converter, as a function of the output voltage V_o and the C_x value

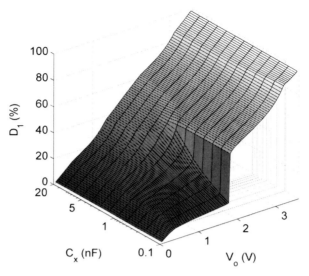

Fig. 5.14 Control signal-to-output voltage transfer function of the 3-level Buck converter, as a function of the output voltage V_o and the C_x value

It is interesting to observe that a low C_x increases the output voltage range in which the converter is DCM operated, and, in turn, the non-linearity of the transfer function (specially for $V_o < V_{bat}/2$).

The transfer function results are compared with those from the classical Buck converter in Fig. 5.15 (the dotted line corresponds to the classical Buck). In this case, just a family of curves with some C_x values is presented for the 3-level converter: 0.1, 1, 3, 5, 10, 20 nF.

From Fig. 5.15 it is observed a more linear behavior for the 3-level Buck converter, specially for higher C_x values; although both of them operate in DCM mode for most of the output voltage range (for the given set of converter characteristics). Despite this fact is not directly related with the feasibility of the converter implementation itself, it should be taken into account when attempting to develop the control scheme.

Figure 5.16 shows the inductor current RMS value resulting from the 3-level converter, as a function of the output voltage and C_x capacitor. The inductor current has been chosen as a comparison index because it becomes a representative index of the RMS current flowing through the rest of the components, which are, in turn, directly related with the corresponding conduction losses.

In Fig. 5.17, the 3-level converter results are compared to the inductor current RMS value of the classical Buck converter. A family of curves are depicted for the 3-level converter for different C_x values (the same ones used in Fig. 5.15).

The comparison in Fig. 5.17 clearly shows that lower conduction losses should be expected for the 3-level converter, since the RMS value of the inductor current is lower, specially in the mid-range of V_o. However, it should be reminded that in the

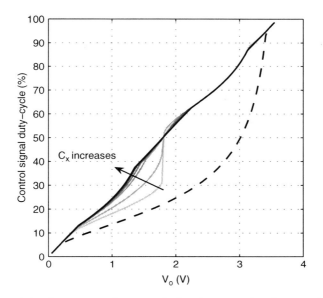

Fig. 5.15 Control signal-to-output voltage transfer function comparison between the 3-level and the classical (*dotted line*) Buck converters. A family of curves with different C_x values is presented for the 3-level converter

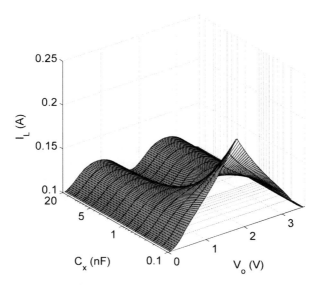

Fig. 5.16 Inductor current RMS value as a function of V_o and C_x, for the 3-level Buck converter

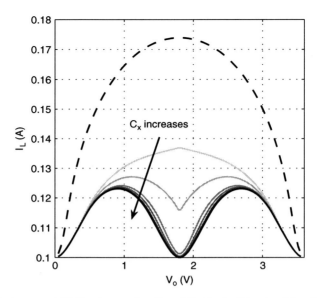

Fig. 5.17 Inductor current RMS value as a function of V_o, for the 3-level and the classical (*dotted line*) Buck converters. A family of curves with different C_x values is presented for the 3-level converter

3-level Buck, inductor current is always flowing through 2 series-connected power MOSFETs, and the C_x capacitor in some states (which could compensate the RMS value reduction).

To continue with the comparison related to the power losses evaluation, maximum switching current (I_{T_1} or I_{T_4} from the analysis from Sect. 5.2.2) is depicted in Fig. 5.18, as a function of V_o and C_x. This magnitude is observed because it is directly related to resistive switching transistors losses.

Figure 5.19 shows the comparison of the results in Fig. 5.18 and the maximum switching current (I_{L_max}) of a classical Buck converter.

Again it seems that lower transistor switching losses should be expected for the 3-level converter, since the state transitions occur in lower inductor current conditions. Nevertheless, it must be taken into account that 4 transistors switch their state along a whole switching period (in front of 2 in the classical Buck converter).

The last issue to compare the performance for both types of converters is their output voltage ripple. The 3-level converter output ripple is depicted in Fig. 5.20, as a function of V_o and C_x.

Figure 5.21 compares these results to the output ripple of the classical Buck converter. Once again, it is observed a better performance for the 3-level converter, that offers a notable reduction of the output ripple, which could result in either an output capacitor decrease (to reduce area occupation), or a switching frequency decrease (to reduce switching losses).

From all the previously presented results, it seems that the 3-level Buck converter could be a reasonable alternative to be integrated, because of its lower output ripple, and possible lower power losses.

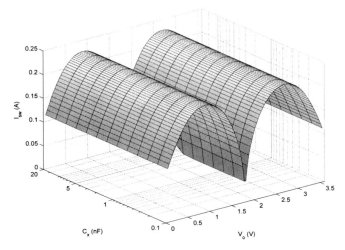

Fig. 5.18 Inductor switching current as a function of V_o and C_x, for the 3-level Buck converter

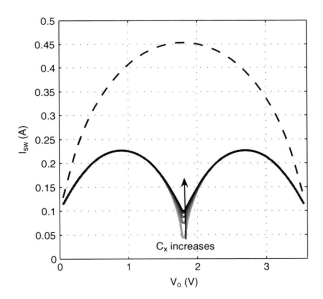

Fig. 5.19 Inductor switching current as a function of V_o, for the 3-level and the classical (*dotted line*) Buck converters. A family of curves with different C_x values is presented for the 3-level converter

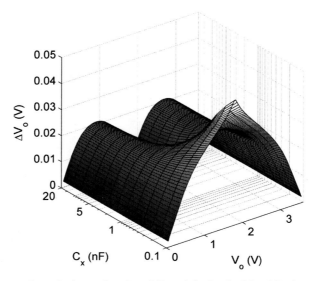

Fig. 5.20 Output voltage ripple as a function of V_o and C_x, for the 3-level Buck converter

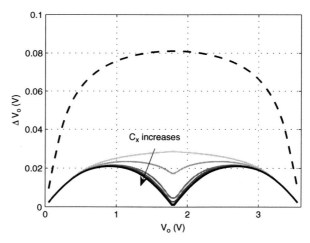

Fig. 5.21 Output voltage ripple as a function of V_o, for the 3-level and the classical (*dotted line*) Buck converters. A family of curves with different C_x values is presented for the 3-level converter

5.3 Self-Driving Scheme

In this section, a particular connection scheme for the power MOSFETs and their drivers is presented to take advantage of the particular structure and operation of the 3-level converter, for a given set of technological parameters of the target microelectronic process.

One of the main issues to be faced when addressing the microelectronic implementation of a switching power converter is the breakdown voltage of the thin gate oxide of the power MOSFETs (which in this work is called V_{max}). This is, for the most recent technologies it is interesting to use the shortest MOSFETs channel to reduce their on-resistance as well as their parasitic capacitances. Unfortunately, the reduction of channel length implies the use of thinner gate oxide dielectrics, that offer lower breakdown voltage (due to the high electric field). In practice, this means that the gate-to-source or drain-to-source voltage excursions must not exceed a ceratin value V_{max} (for reliability purposes). As a consequence, usually input–output transistors are used as power MOSFETs because of their thicker gate dielectric, at the expenses of larger channel lengths.

The herein presented connection scheme, which can be observed in Fig. 5.22, is intended to solve the aforementioned issue to allow the use of core transistors (in case of the considered process: mixed-signal 0.25 μm from UMC).

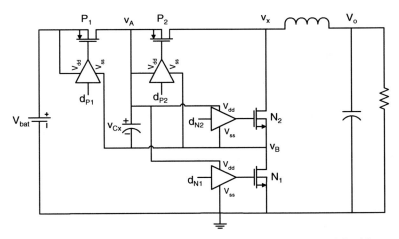

Fig. 5.22 Proposed self-driving scheme to connect both power transistors and the drivers, which reduces the voltage across the power MOSFETs gate dielectric

Provided that the C_x capacitor is large enough, it can be assumed that ΔV_{C_x} throughout the whole switching cycle is insignificant and that V_{C_x} remains constant at the $V_{bat}/2$ value (in our application this means 1.8 V). Hence, $v_A - v_B = V_{bat}/2$, which is an important fact to take into account in the circuit analysis.

It is proposed to use this capacitor as a secondary energy source to supply the drivers, which presents a voltage reduction down to half of the battery voltage by means of a lossless LC mechanism (similar to any other switching converter). It is important to note that the stored charge in the capacitor is much higher than that required by the drivers, and additionally, it is recycled every switching cycle.

The circuit analysis is based on the values shown in Table 5.2.

Table 5.2 v_A and v_B values along the four different converter states (large C_x capacitor is considered)

	T_1	T_2	T_3	T_4
v_A	V_{bat}	$V_{bat}/2$	$V_{bat}/2$	V_{bat}
v_B	$V_{bat}/2$	0	0	$V_{bat}/2$

- The drivers of P_2 and N_2 are completely supplied by the C_x capacitor, which results in a gate-to-source maximum voltage of V_{C_x}. For the considered technology this is low enough to avoid reliability issues due to the gate oxide breakdown voltage. Thus, core transistors (with shorter channel length) can be used, which reduces the power losses.
- Transistors from power drivers can also be minimum channel length, since their voltage supply is half of the input voltage, too. This not only shrinks down the area occupation, but also reduces driver energy consumption since Q_i and Q_{e1} factors become lower.
- P_1 driver must be designed with input–output transistors since during states T_2 and T_3, it is supplied at V_{bat} voltage. However, P_1 transistor gate voltage is $v_B = V_{bat}/2$ during T_1 and T_4, and $v_A = V_{bat}$ during T_2 and T_3, which means that maximum gate-to-source/drain voltage is $V_{bat}/2$ and consequently P_1 can be a core transistor.
- A similar reasoning applies for N_1 and its driver. During T_1 and T_4, the driver is supplied at V_{bat} voltage (which requires the use of input–output transistors to build up the driver); nevertheless, N_1 can be a core transistor since the maximum gate-to-source/drain voltage is $V_{bat}/2$.

To easily extend this reasoning to the DCM operation, the switching sequence for such conduction mode should be considered. Table 5.3 shows the transistors states along a complete switching cycle, for both cases $V_o > V_{bat}/2$ and $V_o < V_{bat}/2$. Attention should be paid to the switching sequence, since it avoids some useless transistor switchings (which would increase power losses). In the table, inactivity states are noted by T_1' and T_3' according to whether they are previous to (in case of $V_o < V_{bat}/2$) or after ($V_o > V_{bat}/2$) the states T_1 and T_3. It is necessary to distinguish both inactivity states since, as exposed on the table, transistors states are different. It is also interesting to note the fact that although they share the name, inactivity states require different switches configuration depending on the V_o value.

Table 5.3 Switching sequence for any of the transistors of a DCM operated 3-level Buck converter (0 stands for 'off-state', whereas 1 stands for 'on-state')

	$V_o < V_{bat}/2$						$V_o > V_{bat}/2$					
	T_1'	T_1	T_2	T_3'	T_3	T_2	T_3'	T_4	T_1	T_1'	T_4	T_3
P_1	0	1	0	0	0	0	0	1	1	1	1	0
P_2	0	0	0	0	1	0	1	1	0	0	1	1
N_1	0	0	1	1	1	1	0	0	0	0	0	1
N_2	1	1	1	0	0	1	0	0	1	0	0	0

Table 5.4 v_A and v_B values along the inactivity converter states, in case of DCM operation (large C_x capacitor is considered)

	$V_o < V_{bat}/2$		$V_o > V_{bat}/2$	
	T_1'	T_3'	T_1'	T_3'
v_A	$V_o + V_{bat}/2$	$V_{bat}/2$	V_{bat}	V_o
v_B	V_o	0	$V_{bat}/2$	$V_o - V_{bat}/2$

Table 5.4 shows the voltage present at nodes A and B during the inactivity states. Again, applying a similar reasoning to the CCM case, it is observed that the proposed connection scheme satisfies that the gate-to-source/drain voltage of power MOSFETs never exceeds the $V_{bat}/2$ value.

A drawback of such proposal is that the on-resistance of a power MOSFETs is reduced by increasing its gate-to-source voltage; hence, a smaller reduction of the on-resistance is expected from the lower V_{on} voltage application. However, global power loss reduction is still expected, since switching losses should be reduced, too.

All the previous explanations are clarified in Fig. 5.23, which presents the drivers voltage supply, as well as the gate-to-source voltage for all the transistors (in case of PMOS transistors P_1 and P_2, source-to-gate voltage is depicted). These waveforms were obtained from transistor-level simulation results, run on CADENCE software.

It is observed that in all of the power MOSFETs, the maximum gate-to-source voltage is roughly $V_{bat}/2$ (which does not exceed the maximum allowed by the technology for core transistors, $V_{max} = 2.5$ V). Additionally, the supply voltage for drivers of P_2 and N_2 transistors is also half of the battery voltage. Only the drivers voltage supply corresponding to P_1 and N_1, reaches the battery voltage during some converter states.

Some oscillations typical at x-node in DCM operation are observed in the P_1 and N_1 drivers voltage supply during the T_1' inactivity state, because N_2 is conducting and v_x oscillations are transferred to v_A and v_B.

In case of a low C_x value (more suitable to microelectronic implementation), an important consideration arises when attempting to implement the herein proposed self-driving scheme to supply the drivers. Since a low C_x value results in a $\Delta V_{C_x} = V_f - V_0$ increase, it must be taken into account that the C_x capacitor voltage must be large enough to let the driver conveniently operate. Therefore, the converter design must consider a minimum V_{C_x} value (V_0) as a constraint in the design space exploration, avoiding those designs that would preclude the drivers operation.

At this point, it is important to comment a very interesting work from Kursun et al. [36], who presented a Buck converter scheme that allows the use of high input voltages in case of integration on low-voltage CMOS processes. In such work, several transistors are series connected and smartly switched to guarantee that gate-to-source/drain/body voltage does not exceed V_{max} in any case. Unfortunately, additional voltage sources are required to generate the necessary intermediate voltage levels.

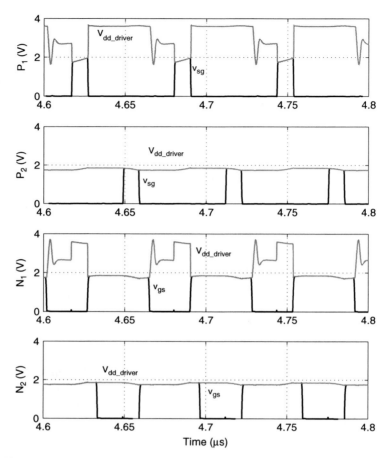

Fig. 5.23 Driver supply voltage and v_{gs} (or v_{sg}) for all power MOSFETs, obtained from transistor-level simulation results

5.4 Specific Components Models and Implementation

5.4.1 C_x Capacitor Implementation

The particular connection of this capacitor (which becomes floating between both nodes A and B) precludes the use of a MOSCAP structure to implement it, since the channel terminal of that structure should be connected to ground.

Inevitably, the microelectronic implementation of the C_x capacitor requires the use of a parallel plates capacitor. Most of the recent mixed-signal technologies (as the considered in this work), have the possibility to use metal layers which are specially close each other (to increase the capacitive density) to implement them.

Hence, in this case, no complex model is required to be included in the design space exploration.

As regards its ESR, it is found that an accurate model would require the use of finite-element simulations since not only it depends on the particular plates shape, but also on the connections placement and size, which in turn rely on the particular layout design of the converter itself and the considered application. For all this reasons, no ESR model has been developed and an arbitrary value (R_{MIM}) for this characteristic will be used in the 3-level converter design space exploration. Associated power losses (P_{C_x}) are then easily computed as the product between the C_x current RMS value and its ESR.

$$P_{C_x} = I_{C_x}^2 R_{MIM} \qquad (5.95)$$

5.4.2 Power MOSFET Energy Losses Evaluation

In this section, the transistors power losses calculation used in case of the 3-level Buck converter design space exploration, are discussed. Conceptually, they are almost coincident with the methodology presented in Sect. 3.4. However, to extend the model to the 3-level converter case, some important additional details are necessary. Hence, in this section just model differences are detailed.

Once again, the same power losses subdivision is used: conduction losses, capacitive switching losses and resistive switching losses.

Because the considered application implies an output voltage of $V_o = 1$ V and an input voltage of $V_{bat} = 3.6$ V, only DCM and CCM operation for the $V_o < V_{bat}/2$ case is exposed. Then, if required, this model can be easily extended to the $V_o > V_{bat}/2$ case by means of similar reasonings.

5.4.2.1 Capacitive Switching Losses E_{sw_C}

In fact, the proposed switching losses model makes its full sense in the case of the 3-level converter, which presents 3 different switching nodes (A, B and x) and 4 power switches connected to them.

First of all, voltage values in the 3 switching nodes, as well as in any power MOSFET gates, are analyzed and summarized in Tables 5.5 and 5.6 for all the state transitions of both operating modes. In these tables, low C_x capacitor is considered to make calculations more accurate, which explains the use of initial (V_0) and final (V_f) C_x capacitor voltages, rather than the $V_{C_x} = V_{bat}/2$ approach (valid for large C_x).

Regarding A, B and x switching nodes, those voltage transitions which are carried out by means of the inductor current are marked by a gray background, to indicate that they are lossless (provided that ideal dead-time is applied).

Then, parasitic capacitances connected to all switching nodes are determined, as shown by expressions (5.96), (5.97) and (5.98).

Table 5.5 Voltage transitions in the switching nodes of the 3-level Buck converter, for both operating modes

DCM

Node	$T_2 \to T_1'$	$T_1' \to T_1$	$T_1 \to T_2$	$T_2 \to T_3'$	$T_3' \to T_3$	$T_3 \to T_2$
v_A	$V_0 \to V_0 + V_o$	$V_0 + V_o \to V_{bat}$	$V_{bat} \to V_f$	$V_f \to V_f$	$V_f \to V_f$	$V_0 \to V_0$
v_B	$0 \to V_o$	$V_o \to V_{bat} - V_0$	$V_{bat} - V_f \to 0$	$0 \to 0$	$0 \to 0$	$0 \to 0$
v_x	$0 \to V_o$	$V_o \to V_{bat} - V_0$	$V_{bat} - V_f \to 0$	$0 \to V_o$	$V_o \to V_f$	$V_0 \to 0$

CCM

Node	$T_2 \to T_1$	$T_1 \to T_2$	$T_2 \to T_3$	$T_3 \to T_2$
v_A	$V_0 \to V_{bat}$	$V_{bat} \to V_f$	$V_f \to V_f$	$V_0 \to V_0$
v_B	$0 \to V_{bat} - V_0$	$V_{bat} - V_f \to 0$	$0 \to 0$	$0 \to 0$
v_x	$0 \to V_{bat} - V_0$	$V_{bat} - V_f \to 0$	$0 \to V_f$	$V_0 \to 0$

Table 5.6 Power MOSFETs gate voltage transitions (v_{gs}), for both operating modes

DCM

Transistor	$T_2 \to T_1'$	$T_1' \to T_1$	$T_1 \to T_2$	$T_2 \to T_3'$	$T_3' \to T_3$	$T_3 \to T_2$
P_1	$0 \to 0$	$0 \to V_0$	$V_f \to 0$	$0 \to 0$	$0 \to 0$	$0 \to 0$
P_2	$0 \to 0$	$0 \to 0$	$0 \to 0$	$0 \to 0$	$0 \to V_f$	$V_0 \to 0$
N_1	$V_0 \to 0$	$0 \to 0$	$0 \to V_f$	$V_f \to V_f$	$V_f \to V_f$	$V_0 \to V_0$
N_2	$V_0 \to V_0$	$V_0 \to V_0$	$V_f \to V_f$	$V_f \to 0$	$0 \to 0$	$0 \to V_0$

CCM

Transistor	$T_2 \to T_1$	$T_1 \to T_2$	$T_2 \to T_3$	$T_3 \to T_2$
P_1	$0 \to V_0$	$V_f \to 0$	$0 \to 0$	$0 \to 0$
P_2	$0 \to 0$	$0 \to 0$	$0 \to V_f$	$V_0 \to 0$
N_1	$V_0 \to 0$	$0 \to V_f$	$V_f \to V_f$	$V_0 \to V_0$
N_2	$V_0 \to V_0$	$V_f \to V_f$	$V_f \to 0$	$0 \to V_0$

$$C_A = C_{ddP1} + C_{jdP1} + C_{sgP2} + C_{sdP2} \tag{5.96}$$

$$C_B = C_{ddN1} + C_{jdN1} + C_{sgN2} + C_{sdN2} \tag{5.97}$$

$$C_{V_x} = C_{ddP2} + C_{jdP2} + C_{ddN2} + C_{jdN2} + C_{PAD} \tag{5.98}$$

It must be observed that C_A and C_B parasitic capacitors do not include the C_x capacitor. This is because this capacitor does not change its voltage due to the converter state changes. It just acts as an electrical link between A and B nodes; this is, any voltage change on any of both nodes is reproduced at the other capacitor terminal. Only the inductor current during T_1 and T_3 states changes the V_{C_x} voltage, or the supply current of the drivers.

Then, tables for all the required parasitic capacitances are obtained from transistor-level simulations, as explained in Sect. 3.4.2.1. Nevertheless, if the type of transistors is changed (e.g. core transistors instead of input–output transistors) new tables must be obtained.

Finally, energy losses related to the charge and discharge process of all the parasitic capacitors at the switching nodes are evaluated (considering that they are nonlinear).

5.4.2.2 Resistive Switching Losses E_{sw_R}

Resistive switching losses are evaluated as in the case of the classical Buck converter. In Table 5.7 the switching currents for any transistor and any converter state transition are shown (for CCM and DCM). Additionally, gray background is used to mark those cells corresponding to situations in which no resistive switching losses occur because either switching current is zero or the particular transistor does not switch in that transition (represented by a black dot).

Table 5.7 Power MOSFETs switching current values of the 3-level Buck converter, for both operating modes (the black dots indicate those situations where the transistor does not switch)

DCM						
Transistor	$T_2 \to T_1'$	$T_1' \to T_1$	$T_1 \to T_2$	$T_2 \to T_3'$	$T_3' \to T_3$	$T_3 \to T_2$
P_1	•	0	I_{T_1}	•	•	•
P_2	•	•	I_{T_1}	•	0	I_{T_1}
N_1	0	•	I_{T_1}	•	•	I_{T_1}
N_2	•	•	•	0	•	I_{T_1}

CCM			
Transistor	$T_2 \to T_1$	$T_1 \to T_2$	$T_2 \to T_3$	$T_3 \to T_2$
P_1	I_0	I_{T_1}	•	•
P_2	•	•	I_0	I_{T_1}
N_1	I_0	I_{T_1}	•	•
N_2	•	•	I_0	I_{T_1}

The corresponding values for I_{T_1} and I_0 can be obtained from the expressions in Sect. 5.2.

Then, using expression (3.80), resistive switching losses can be computed. t_{sw} values must be obtained from the tapered buffers design for any of the power transistors, and the required R_{TRT} values are extracted from tables obtained from transistor-level simulations.

It is interesting to note at this point that, when designing the drivers, the considered voltage supply is no longer the battery voltage if the self-driving scheme presented in Sect. 5.3 is implemented. Instead of this, the drivers voltage supply is set to $V_{bat}/2$, as a coarse approach.

5.4.2.3 Conduction Losses

Although the concept used for transistor conduction losses evaluation is the same as in Sect. 3.4.3, important modifications must be included in the 3-level converter case, if a low C_x capacitor is implemented.

The reason for this is that the on-state gate-to-source voltage (V_{on}) is not constant along the complete switching cycle, for all the power transistors, because the C_x voltage changes from V_0 to V_f.

During the T_2 state, N_1 and N_2 v_{gs} voltage is constant, but it is V_0 before the T_1 state and V_f after it. Furthermore, during T_1 state, P_1 and N_2 applied gate voltage continuously changes from V_0 to V_f; and this trend is reversed along the T_3 state for the applied voltage to P_2 and N_1 gates ($v_{C_x} = V_f \rightarrow V_0$). This information is summarized in Table 5.8 (obviously, in this case, no necessary distinction is needed between DCM and CCM).

Table 5.8 Applied gate voltages at the 3-level converter transistors, during their on-state (a black dot indicates that the particular transistor is off)

Transistor	T_2	T_1	T_2	T_3
P_1	●	$V_0 \rightarrow V_f$	●	●
P_2	●	●	●	$V_f \rightarrow V_0$
N_1	V_0	●	V_f	$V_f \rightarrow V_0$
N_2	V_0	$V_0 \rightarrow V_f$	V_f	●

All this voltage changes imply that the on-resistance of the conducting transistors changes as well. Thus, the RMS value of the current referred to a complete switching cycle can not be used directly to evaluate the conduction losses.

As a consequence, different solutions are proposed for the different situations.

T_2 state

Conduction losses due to transistors current flowing during the T_2 state are computed as the product between the RMS value I_{LT_2RMS} (obtained in Sect. 5.2) and the sum of the N_1 and N_2 on-resistances, corresponding to V_0 and V_f, as V_{on}. Since energy computation is preferred, the whole expression is divided by the switching frequency (f_s).

$$E_{cond\_T2} = \frac{I_{LT_2RMS}^2}{f_s}(R_{N1\_V_0} + R_{N2\_V_0} + R_{N1\_V_f} + R_{N2\_V_f}) \qquad (5.99)$$

T_1 state

In case of conduction losses corresponding to the T_1 state, a more complex evaluation procedure is proposed, since a variable on-resistance must be taken into account for the conducting transistors. Consequently, a staggered approach is proposed to model the on-resistance (see Fig. 5.24), and the posterior computation is similar to the resistive switching losses calculation. To obtain the different resistance values of the staggered approach, a linear evolution for the v_{gs} voltage is supposed and a $R = f(V_{gs})$ table is used (since the channel resistance is not a linear function of the applied gate voltage, as observed in Fig. 3.59).

Then, energy losses are computed as follows:

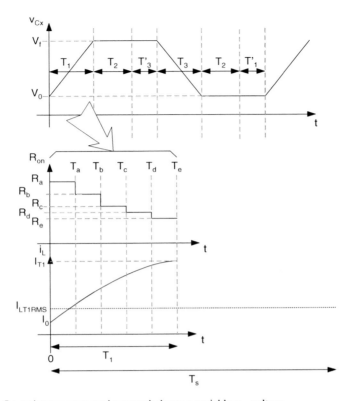

Fig. 5.24 On-resistance staggered approach due to a variable v_{gs} voltage

$$E_{cond\_T1} = R_a \int_0^{T_a} i_L^2(t)\mathrm{d}t + R_b \int_{T_a}^{T_b} i_L^2(t)\mathrm{d}t + R_c \int_{T_b}^{T_c} i_L^2(t)\mathrm{d}t +$$

$$+ R_d \int_{T_c}^{T_d} i_L^2(t)\mathrm{d}t + R_e \int_{T_d}^{T_e} i_L^2(t)\mathrm{d}t \qquad (5.100)$$

This allows to compute the energy corresponding to any piece of the R_{on} staggered function, and finally all the partial calculations can be accumulated.

$$E_{cond\_T1} = \sum_i \frac{I_{LT_1RMS}^2 R_i}{n_R f_s} = \frac{I_{LT_1RMS}^2}{n_R f_s} \sum_i R_i = \frac{I_{LT_1RMS}^2}{n_R f_s} \sum_i (R_{iN} + R_{iP})$$

$$(5.101)$$

where R_i is the ith subdivision of the total on-resistance (i.e. the sum of the resistance corresponding to the NMOS and PMOS series connected transistors); and n_R is the number of subdivisions used in the staggered approach (5, in the example of Fig. 5.24).

T₃ state

The conduction losses calculations corresponding to the T_3 state are carried out in an analogous way to the T_1 state counterpart.

Just to finish the driver design and losses calculation for the 3-level converter, some considerations regarding the design procedure are discussed in the following.

It is interesting to mention that the optimum transistors channel width is concurrently found by means of genetic algorithms, in the same way as for the classical Buck converter. Here, it is specially indicated the use of such optimization method, since the existence of four different variables (W_{P_1}, W_{P_2}, W_{N_1} and W_{N_2}) would make it impractical to explore the whole transistors design space (for any point of the global design space exploration), because of the huge number of possible designs. Again, total power losses (switching and conduction) are added and the optimum design is established as the one that produces the lowest power losses, disregarding occupied area considerations (since it is expected to be insignificant in front of the whole power converter area).

In addition to this, 2 main constraints are applied to any particular guess: a maximum propagation delay on the driver design (t_{d_max}), and a minimum C_x capacitor voltage (since it is necessary to supply the power drivers).

Chapter 6
3-Level Buck Converter Design Space Exploration Results

Abstract Having presented the theoretic operation of the 3-level Buck converter, and the expected benefits that would result from its application as a low-power fully integrated step-down converter, the corresponding design space exploration is presented in this chapter. First, the same output ripple constraint as in case of the classical Buck converter is applied; and the results show an expansion of the possible designs space, due to its relatively lower output ripple. In this sense, an optimized design with higher power efficiency and more reduced area is obtained. However, a new design space constraint is presented from the requirement to provide a wide output current range. This new context demands some kind of switching frequency modulation to keep an acceptable power efficiency when very low output current is provided. Consequently, a proportional modulation (Arbetter and Maksimovic, *Power Electronics Specialists Conference, 1997, PESC'97 Record, 28th Annual IEEE*, St. Louis, MO, June 1997) of the switching frequency as a function of the output current (provided that the converter is DCM operated), is proposed. The main drawback of this proposal is an output ripple increase as both I_o and f_s become lower. Therefore, the new constraint requires to keep the output ripple below 50 mV in the output current range of 5 mA $\rightarrow$ 100 mA. Obviously, this results in a stronger reduction of the design space, and another optimized design is found providing poorer efficiency and bigger area occupancy (which is still better than the classical Buck counterpart). Finally, a suboptimum design selected, to be implemented in the chip to be tested, is presented because of practical considerations about the feasibility of the required control circuitry.

6.1 Design Space Exploration Results for the $I_o = 100\,\text{mA}$ Case

The design space exploration corresponding to the 3-level converter will be carried out in a similar way than the corresponding to the classical Buck converter. Nevertheless, here, the presence of the C_x capacitor adds a new dimension to be explored, since its value has a great impact on important converter magnitudes such as the RMS inductor current value (related to power losses), or the output voltage ripple (that strongly constrains the possible design space) and, obviously, the occupied silicon area. Therefore a 4 dimensional design space exploration should be carried,

G. Villar Piqué, E. Alarcón, *CMOS Integrated Switching Power Converters*,
DOI 10.1007/978-1-4419-8843-0_6, © Springer Science+Business Media, LLC 2011

which implies an overwhelming computation complexity. Even worse, the qualitative interpretation of the results could become really complicated (since a four variables function should be represented).

However, in this work, an advantageous intersection of a particular implementation and a technical possibility (provided by the selected microelectronic process) are used to simplify this issue.

The main fact beyond this proposal is related to both converter capacitors (C_x and C_o) implementation. In Sect. 3.2, a matrix structure was presented to implement the output capacitor by the interconnection of multiple MOS capacitors. Such structure uses the bottom 3 metal layers. On the other hand, as previously commented, the electrical requirements of the C_x capacitor preclude the use of a MOSCAP structure to implement it, thereby leaving the use of a flying *Metal-Insulator-Metal* (MIM) as the only feasible possibility. In the selected process, this kind of metal capacitor is implemented by means of the fourth and fifth metal layers, and, additionally, an special metal layer between both of them, which is extremely close above the Metal-4 layer.

Hence, it is proposed to place the C_x capacitor on the output capacitor, occupying the same area. In the following a list of considerations and benefits arisen from this proposal, is presented.

- In the design space exploration, C_o values are swept. The resulting area occupancy from the C_o capacitor is used to determine the C_x value, by means of the C_{MIM} capacitive density (which in case of the UMC 0.25 μm process is $1\ fF/\mu m^2$). This way any single value of the output capacitor is linked to a single value of C_x, which implies a reduction of the design variables number from 4 to 3. Thus, the design space exploration can be carried out and analyzed as in the classical Buck converter (Chap. 4).
- This simplification makes sense because, as observed in Chap. 5, an increase in the C_x value, results in lower output ripple and power losses (because of lower inductor current RMS value). Then, if C_x is placed on the C_o, and it is supposed that the later occupies more silicon area it is interesting to take advantage of the unused area in the top metal layers, and increase the C_x value. On the contrary, if C_x is supposed to occupy more area than C_o (because their values are not linked), it is interesting to increase the output capacitor so as to reduce the output ripple.
- An obvious benefit that stems from the 'sandwich' placement of both capacitors is the reduction of the total occupied area (similar to the case of the inductor bonding wire implementation).
- A possible drawback of this proposal is that an important parasitic capacitor appears between one of the C_x capacitor plates and the ground node (since, in the MOSCAP structure, the Metal-3 layer is used as the ground terminal). Although this effect has not been studied in this work, it could be interesting to investigate it as a future research line.

Other considerations (no related to the C_x capacitor implementation) to take into account before carrying out the design space exploration are listed in the following:

- As in case of the classical Buck converter, the *skin-effect* is also considered in the inductor ESR calculations (see Chap. 4).
- Since more switching activity is required for any switching cycle of the 3-level converter (at least 4 different states are changed), the maximum allowed propagation delay of any of the tapered buffers, is constrained down to $t_{d\_max} = 1$ ns.
- Power drivers are considered to be supplied at the $V_{bat}/2$ voltage, because of the self-driving scheme, presented in Chap. 5.
- Since the power drivers are supplied from the C_x capacitor, a minimum value of 1.2 V for the V_{C_x} voltage is used as a design space constraint.
- Although P_1 and N_1 transistors are implemented by core transistors ($L_{ch} = 0.25\,\mu\text{m}$), their drivers are implemented by means of input – output transistors ($L_{ch} = 0.35\,\mu\text{m}$).
- Both P_2 and N_2 power transistors, as well as their corresponding power drivers, are implemented by means of core transistors.
- Both PMOS power switches (P_1 and P_2) are implemented by means of Low-V_t transistors, since, in this particular technological process, they offer lower resistance at the expenses of an slightly increase of their channel length ($L_{ch} = 0.3\,\mu\text{m}$, although being core transistors).

As in case of the classical Buck converter, the most important application and technical parameters are summarized in Table 6.1.

Figures 6.1, 6.2 and 6.3 show the design exploration results in terms of power efficiency. As in Chap. 4, white squares are used to depict those design configurations that are excluded from the possible results because they yield an output ripple higher than the maximum allowed. To depict possible designs, gray-scale zones are used, where the darker the area, the higher the efficiency. Additionally, black diamonds are used to mark those design configurations that result in CCM operation. Finally, a black dot marks the maximum efficiency configuration.

In the upper right corner, the design variables values corresponding to the black-dot design are annotated, whereas the corresponding specific design data are presented in the upper left corner.

An interesting particularity of the proposed representation is that for low f_s values, there are zones of the (L, C_o) plane that do not present any square. These correspond to designs that where rejected not for the output voltage ripple constraint, but for the minimum $V_{C_x} > 1.2$ V constraint (since it is considered that the converter would not work under that condition). Consequently, these 'forbidden' zones are found for low C_o values, which imply low C_x values. It must be remembered from Sect. 5.2, that Δv_{C_x} is inversely proportional to C_x capacitor.

Although, the power efficiency evolution is similar to the classical Buck converter case, some important differences appear.

First, the vertical scale of the 3-dimensional surfaces has changed from $47.75\% \rightarrow 62.5$ to $50.75\% \rightarrow 70.25\%$, which implies a general increase of the power efficiency.

Table 6.1 Design parameters and technical information used in the Buck converter design space exploration (UMC 0.25 μm mixed-signal process)

Application parameters			
Battery voltage (V_{bat})			3.6 V
Output voltage (V_o)			1 V
Output current (I_o)			100 mA
Maximum output voltage ripple (ΔV_o)			50 mV
Inductor design			
r	12.5 μm	p	50 μm
σ	41×10^6 S/m	R_{bp}	13 mΩ
$\mu_{r\_conductor}$	0.99996	μ_0	$4\pi \times 10^{-7}$ H/m
μ_{r_media}	1	L tolerance	3%
γ_{LA}	1	γ_{LR}	10
Output capacitor design (Core transistors)			
V_{gs}	1 V	C_{ox}	6.27 fF/μm^2
μ_N	376 cm^2/(sV)	V_{term}	25 mV
α	1/12	γ	12
a	0.8 μm	b	0.8 μm
R_{poly}	2.5 Ω	R_{MET1}	53 mΩ
R_{MET2}	53 mΩ	R_{MET3}	53 mΩ
δ_{cont}	1/0.72 cont./μm	δ_{via1}	1/0.76 Via1/μm
δ_{via2}	1/0.76 Via2/μm	R_{cont}	5 Ω/cont.
R_{via1}	3.5 Ω/via1	R_{via2}	3.5 Ω/via2
V_{TN}	0.17 V		
C_x capacitor design (Metal-Insulator-Metal)			
C_{MIM}	1 fF/μm^2	R_{MIM}	0.1 Ω
N_1 and P_1 drivers desgin (IO transistors)			
L_{min}	0.35 μm	W_{min}	0.3 μm
C_{PAD}	1 pF	t_{d_max}	1 ns
t_{di}	Table	t_{de1}	Table
t_{fri}	Table	t_{fre1}	Table
Q_i	Table	Q_{e1}	Table
All power transistors and N_2 and P_2 drivers design (IO transistors)			
L_{min}	0.25 μm	W_{min}	0.3 μm
C_{PAD}	1 pF	t_{d_max}	1 ns
t_{di}	Table	t_{de1}	Table
t_{fri}	Table	t_{fre1}	Table
Q_i	Table	Q_{e1}	Table

Also, it is observed that even for $f_s = 10$ MHz possible designs appear which is due to the lower output voltage ripple produced by the 3-level converter, expanding the area of the possible design sets.

Again, it is observed that in all the frames the highest power efficiency is obtained in for DCM operated design, although it is on the edge between both operating modes. The reason for this is that this placement produces an interesting trade-off between the soft-switching conditions arisen from the DCM operation and the lower current RMS values, obtained from the CCM operation (mainly, due to higher

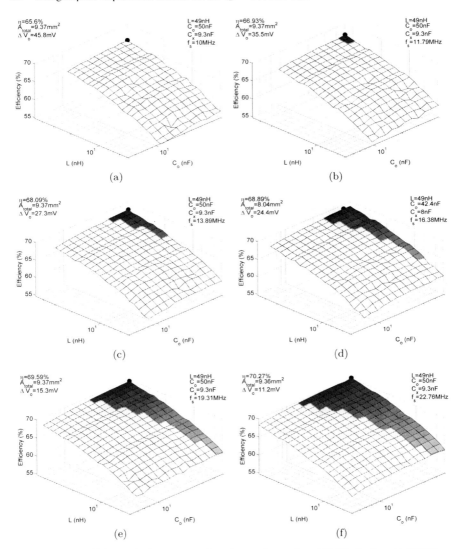

Fig. 6.1 Power efficiency design space exploration results. **a** $f_s = 10$ MHz, **b** $f_s = 11.79$ MHz, **c** $f_s = 13.89$ MHz, **d** $f_s = 16.38$ MHz, **e** $f_s = 19.31$ MHz, **f** $f_s = 22.76$ MHz

inductance values). It is interesting to note that power efficiency is clearly an increasing function of the L value, (except in the CCM zone), while it becomes less dependent of the output capacitor value, since this only produces slight changes on its ESR and the RMS value of the capacitor current is relatively low. A greater impact could result from the C_x capacitor ESR value, because higher amounts of current flow through it, but, in this work, its ESR has been considered constant.

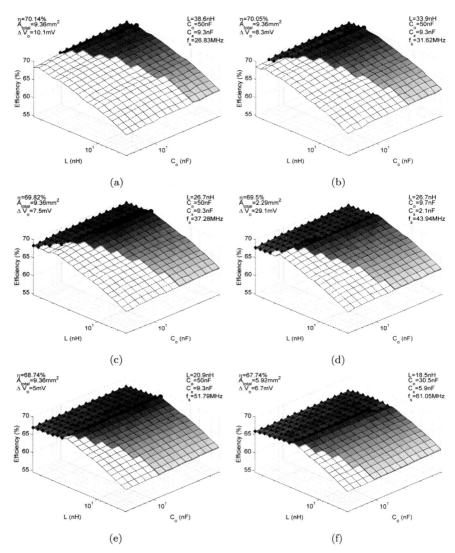

Fig. 6.2 Power efficiency design space exploration results. **a** $f_s = 26.83$ MHz, **b** $f_s = 31.62$ MHz, **c** $f_s = 37.28$ MHz, **d** $f_s = 43.94$ MHz, **e** $f_s = 51.79$ MHz, **f** $f_s = 61.05$ MHz

However, for low C_o values an efficiency reduction is observed because of the resulting lower C_x value, that yields a higher RMS value of the inductor current.

In this perspective, power efficiency also presents a maximum as a function of the switching frequency (Fig. 6.1f, $f_s = 22.76$ MHz $\rightarrow$ $\eta = 70.27\%$), as a consequence of the more proper balance between the switching and the conduction losses. At this point, it should be observed that an increase in the maximum power efficiency from 62.38% (Fig. 4.2a) up to 70.27% has been achieved by changing

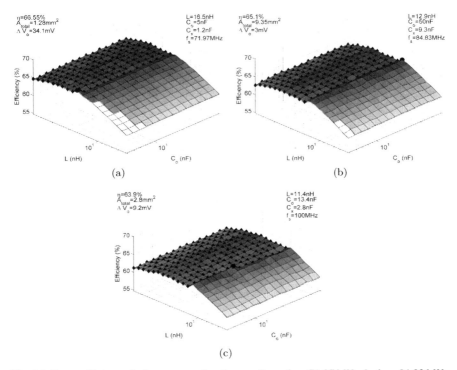

Fig. 6.3 Power efficiency design space exploration results. **a** $f_s = 71.97\,\text{MHz}$, **b** $f_s = 84.83\,\text{MHz}$, **c** $f_s = 100\,\text{MHz}$

the converter topology. In contrast to the classical Buck case, the maximum power efficiency appears for a much lower switching frequency (approximately half the value). However, it should be taken into account that the whole switching cycle of the 3-level converter implies the double state changes (this is, the double switching activity).

Total occupied area results are depicted in Figs. 6.4, 6.5 and 6.6, where similar representation criteria as with the power efficiency, were used. Here, the area has been computed as the maximum of the inductor area or the sum of the output capacitor area plus the power switches and drivers area. The corresponding expression from Chap. 4 is repeated here, for the sake of clarity.

$$A_{total} = max[A_L \quad (A_{C_o} + A_{PMOS} + A_{NMOS} + A_{Pdriver} + A_{Ndriver})] \qquad (6.1)$$

The results are very similar to the obtained for the classical Buck converter, since the same values were used to sweep the L and C_o design variables (which are responsible for almost the total occupied area). In this case, the proposed placement of the C_x capacitor above the output capacitor does not change the occupied area evolution. However, it is also noted the increase of the design space (referred to the 3 design variables) due to the lower output ripple that results from the 3-level

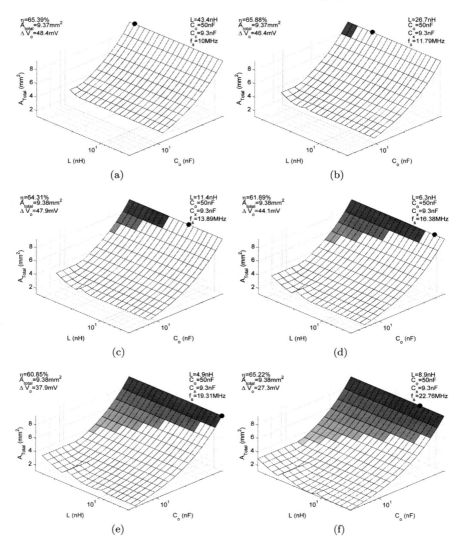

Fig. 6.4 Total occupied area design space exploration results. **a** $f_s = 10$ MHz, **b** $f_s = 11.79$ MHz, **c** $f_s = 13.89$ MHz, **d** $f_s = 16.38$ MHz, **e** $f_s = 19.31$ MHz, **f** $f_s = 22.76$ MHz

converter operation. Therefore, when analyzing these area results, the same reasonings for the classical Buck can be applied here.

In order to evaluate the global performance of the designed converter, a merit figure that considers both occupied area and power efficiency is used. To carry out an homogeneous comparison with classical Buck converter, the same merit figure definition has been used here (expression (6.2) remembers it).

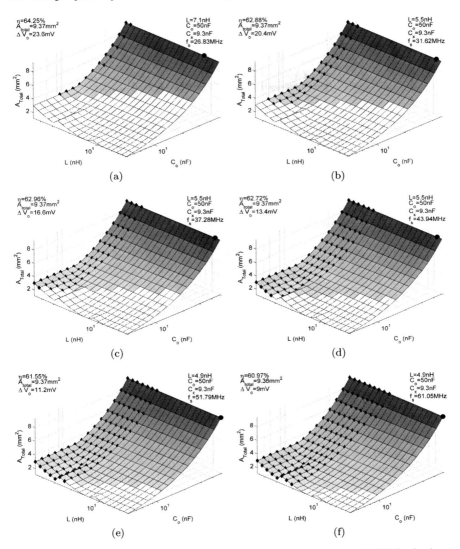

Fig. 6.5 Total occupied area design space exploration results. **a** $f_s = 26.83$ MHz, **b** $f_s = 31.62$ MHz, **c** $f_s = 37.28$ MHz, **d** $f_s = 43.94$ MHz, **e** $f_s = 51.79$ MHz, **f** $f_s = 61.05$ MHz

$$\Gamma = \frac{(\eta - \eta_{min})^2}{A_{total}} \qquad (6.2)$$

Although similar criteria to the power efficiency and occupied area cases have been applied to present the merit figure evaluation results, it must be taken into account that L and C_o axis have been reversed, resulting in a turn of $180\,^{\circ}$ in the perspective view of the 3-dimensional surfaces (Figs. 6.7, 6.8 and 6.9).

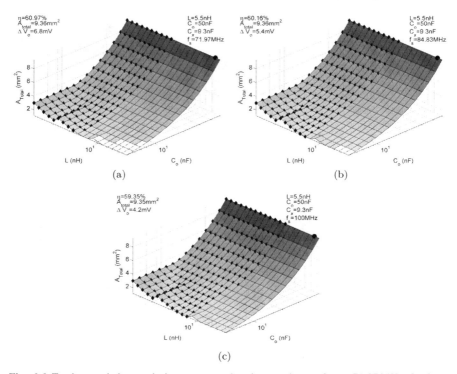

Fig. 6.6 Total occupied area design space exploration results. **a** $f_s = 71.97$ MHz, **b** $f_s = 84.83$ MHz, **c** $f_s = 100$ MHz

As a qualitative approach, it is observed a relative increase of vertical axis ticks values which imply a better performance, since the merit figure definition is coincident for both converters.

Once again, in most of the frames the maximum merit figure value is produced by designs that are on the border of the 'forbidden' zone, because they provide the lowest output capacitor (which implies the lowest area occupancy) that produces an acceptable output ripple. Nevertheless, for high frequency values, the maximum merit figure design is no more on that border, since the lowest considered capacitor is achieved and then the inductor area or the power efficiency take their effect.

The impact of the merit figure evaluation in the design selection can be observed by comparing the efficiency and the total area of the maximum power efficiency design and the maximum merit figure design. While the first yields a power efficiency of 70.27% and $A_{total} = 9.36$ mm^2, the later results in 69.33% and 1.81 mm^2, respectively. Thus, an important reduction of the occupied area is obtained at the expenses of an slight decrease of the power efficiency.

The maximum merit figure value is obtained in Fig. 6.2c. This design yields a power efficiency of 69.33% and a occupied area of 1.81 mm^2, which are much

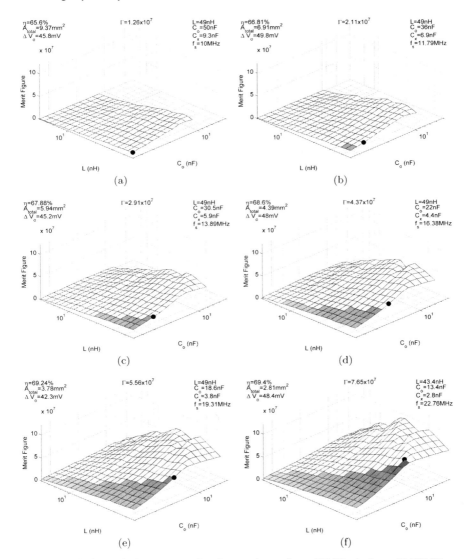

Fig. 6.7 Merit figure design space exploration results. **a** $f_s = 10\,\text{MHz}$, **b** $f_s = 11.79\,\text{MHz}$, **c** $f_s = 13.89\,\text{MHz}$, **d** $f_s = 16.38\,\text{MHz}$, **e** $f_s = 19.31\,\text{MHz}$, **f** $f_s = 22.76\,\text{MHz}$

better than the same results for the selected design of the classical Buck converter: $\eta = 61.02\%$ and $A_{total} = 2.29\,\text{mm}^2$.

It is interesting to observe the evolution of the maximum merit figure value of each frame, as a function of the switching frequency. In Fig. 6.10, this evolution is compared to the previously presented for the classical Buck converter.

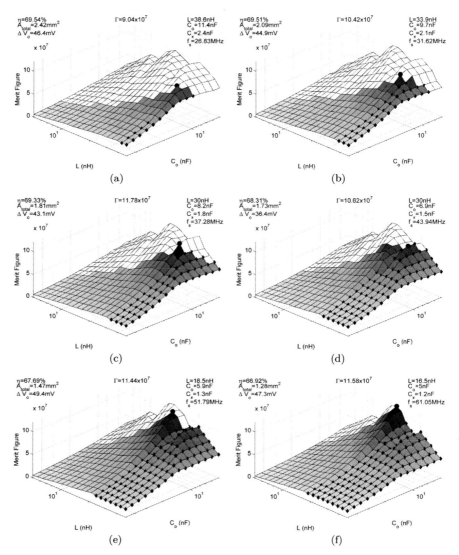

Fig. 6.8 Merit figure design space exploration results. **a** $f_s = 26.83$ MHz, **b** $f_s = 31.62$ MHz, **c** $f_s = 37.28$ MHz, **d** $f_s = 43.94$ MHz, **e** $f_s = 51.79$ MHz, **f** $f_s = 61.05$ MHz

As expected, for almost all the switching frequency values here considered the merit figure takes a higher value for the 3-level converter. The possible low frequency designs also appear in the comparison, since only 3-level converter designs are possible below the 20 MHz limit due to the output ripple constraint.

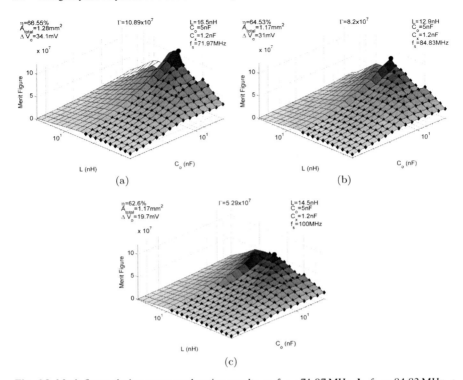

Fig. 6.9 Merit figure design space exploration results. **a** $f_s = 71.97\,\text{MHz}$, **b** $f_s = 84.83\,\text{MHz}$, **c** $f_s = 100\,\text{MHz}$

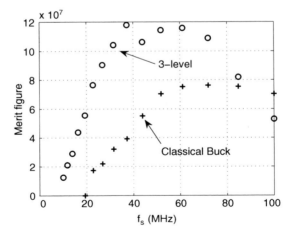

Fig. 6.10 Maximum merit figure value as a function of the switching frequency, for the classical and the 3-level converters

6.2 Design Space Exploration Results for a I_o Wide Range

From the previous reasoning, it could seem that a converter design offering acceptable results in terms of power efficiency and occupied area, while keeping an output ripple lower enough for the considered application, have been achieved. However, at this point an additional and important consideration should be taken into account.

The main issue is to consider an output load that changes its current consumption for a wide range, from its nominal value (considered as the maximum), down to the microamperes range. This is a very common situation when supplying digital cores that change their working mode according to their computation needs (e.g. full-processing and sleep modes), which involves a significant current consumption change, of several orders of magnitude.

If the previous assumption is considered, several important implications should be taken into account. First of all, DCM operating mode will the unavoidable at very low output current values. On the other hand, high switching frequency could result in very low power efficiency when supplying low output current, due to switching losses.

Therefore, some kind of switching frequency modulation should be considered as the output current changes, since it is the only dynamic design variable for a given converter design (i.e. having determined all the static design variables, such as (L, C_o, C_x)).

A complete dissertation about the optimum switching frequency modulation is presented in Appendix C. The main conclusion is that, provided that the converter is DCM operated, a linear switching frequency modulation keeps the power efficiency constant as the output current changes.

$$f_s = kI_o \tag{6.3}$$

As observed in expression (6.3), the switching frequency is proposed to be proportional to the output current, since when no output current is supplied any activity is not required in the converter. Thus, the singular point of $I_o = 0\,A \rightarrow f_s = 0\,Hz$ is consistent with this reasoning.

In order to keep the optimum power efficiency, the slope of the linear function (k) should be stated as the ratio between the switching frequency of the selected design and maximum output current of the considered application.

$$f_s = \frac{f_{s\_opt}}{I_{o\_max}} I_o \tag{6.4}$$

Figure 6.11 shows the evolution of the power efficiency for a wide range of the output load current, for both cases: keeping constant the switching frequency or applying the linear modulation. The optimum switching frequency ($f_{s\_opt} = 37.28\,MHz$) is obtained from the results of Fig. 6.2c, while the maximum output current value is considered to be $100\,mA$.

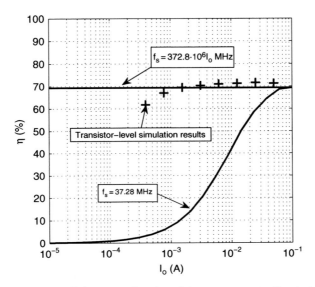

Fig. 6.11 3-level converter efficiency as a function of the output current. Constant switching frequency against linear modulation are compared

From these results, it is observed that if a variable current load must be supplied the linear f_s modulation as the output current changes becomes fundamental.

In this case, model results are contrasted against transistor-level simulation results, for the case of applying the proportional f_s modulation. Good matching is observed between simulation measurements and theoretical power efficiency for more than a decade of the I_o values. Nevertheless, for very low output current values, power efficiency starts to decrease because of the leakage current through the power MOS switches, which becomes more important as the output power is reduced.

Although this f_s modulation contributes to the fundamental objective of keeping the power efficiency as high as possible, a new important drawback arises: the output voltage ripple is increased as the I_o is reduced. This effect appears because of the switching frequency reduction, as a consequence of the proposed modulation.

A qualitative approach to this phenomenon can be easily explained:

1. The main consequence of the proportionality between the f_s and the output current is that all the inductor current pulses (it must remembered that the converter must be DCM operated to apply the linear modulation) keep exactly the same shape, and just their frequency is changed (for further details on this issue refer to the Appendix C).
2. In case of an extremely low output current, the entire inductor current pulse is stored in the output capacitor since almost no output current is substracted from the capacitor ($i_{C_o} = i_L - I_o$).

3. This generates an increase of the Δv_o, that decreases as the output current becomes higher, since the total charge stored in the output capacitor along the whole switching cycle becomes lower.

This behavior can be observed in the Fig. 6.12, where the output ripple evolution as a function of the output current is depicted. The case of the switching frequency remaining constant is compared to the resulting from the (6.3) application. For the first case, the output ripple is reduced as I_o becomes lower because the inductor current pulses become smaller, resulting in little charge to be stored in the output capacitor (which reduces the Δv_o magnitude).

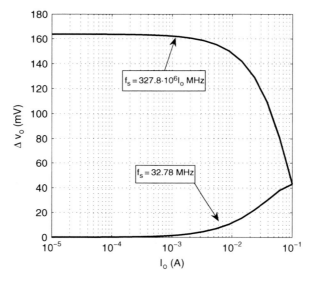

Fig. 6.12 3-level converter output ripple as a function of the output current. Constant switching frequency against linear modulation are compared

Conversely, the output ripple is increased as I_o is reduced if the linear switching frequency modulation is applied, as expected from the previous qualitative reasoning.

From the comparison results of Figs. 6.11 and 6.12, it is concluded that keeping the switching frequency constant reduces the output ripple for low output current, whereas the power efficiency is shrunk almost down to zero. In this respect, in this work, keeping an acceptable power efficiency is considered to be more important than reducing the output ripple far below the value required by the output load.

From the curve of Fig. 6.12, it is observed that the most of the ripple increase it produced in the miliamperes range. Then, as the output current is further reduced, only an slight output ripple increase is produced. Moreover, it is common that digital cores became more tolerant to the voltage supply ripple when they remain in power-saving modes such as sleep mode.

In conclusion, it is proposed to obtain a design that assures that the output ripple is below the 50 mV constraint in the output current range of 5 mA $\rightarrow$ 100 mA.

Therefore, in front of this new constraint, obtained design space exploration results are discarded, and design space should be reexplored constraining even more the design space, by those designs that provide an output ripple lower than 50 mV in the whole range of the considered output current values (provided that the $f_s = kI_o$ modulation is applied to maintain an acceptable power efficiency).

It is very important to note that, under this new application conditions, all the configuration sets that yield CCM operation are precluded, because it would be unavoidable to change their operating mode for very low I_o values (and unexpected

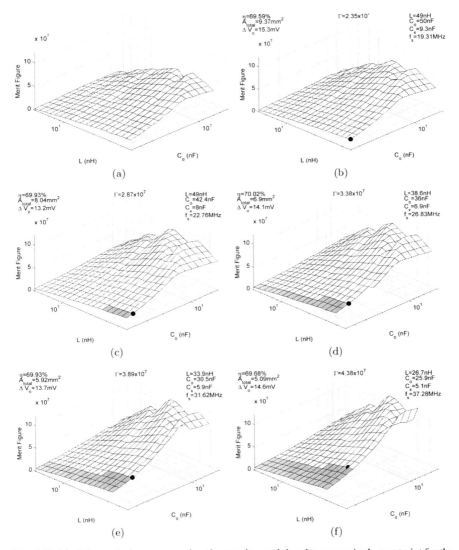

Fig. 6.13 Merit figure design space exploration results, applying the output ripple constraint for the 5 mA $\rightarrow$ 100 mA output current range. **a** $f_s = 16.38$ MHz, **b** $f_s = 19.31$ MHz, **c** $f_s = 22.76$ MHz, **d** $f_s = 26.83$ MHz, **e** $f_s = 31.62$ MHz, **f** $f_s = 37.28$ MHz

power efficiency would result). Furthermore, a change in the converter operating mode could result in important control loop stability issues.

Figures 6.13 and 6.14 expose the new exploration results in terms of the merit figure evaluation, having applied the constraints on the converter design to be DCM operated (at $I_o = 100\,\text{mA}$), and to provide an output ripple lower than 50 mV for $I_o = 5\,\text{mA} \rightarrow 100\,\text{mA}$ range (considering the proportional modulation of the

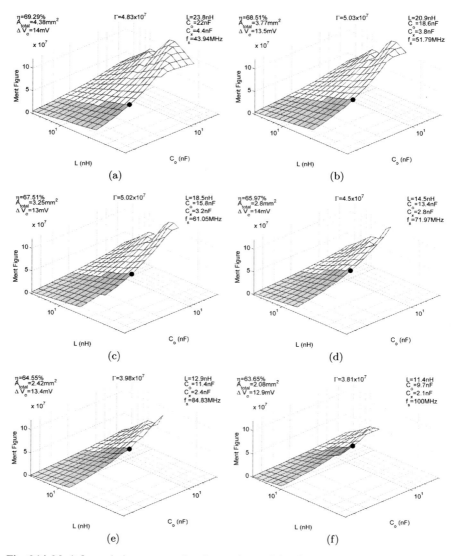

Fig. 6.14 Merit figure design space exploration results, applying the output ripple constraint for the 5 mA → 100 mA output current range. **a** $f_s = 43.94\,\text{MHz}$, **b** $f_s = 51.79\,\text{MHz}$, **c** $f_s = 61.05\,\text{MHz}$, **d** $f_s = 71.97\,\text{MHz}$, **e** $f_s = 84.83\,\text{MHz}$, **f** $f_s = 100\,\text{MHz}$

switching frequency). Here, white squares are used to marked those designs discarded because of the output ripple constraint, while CCM operated designs have not been depicted.

From these results, it observed that the maximum merit figure value is obtained from the design of Fig. 6.14b, that yields a power efficiency of 68.51% and occupies an area of 3.77 mm^2. In Table 6.2 the most important data about this design is summarized.

And Fig. 6.15 shows the output voltage ripple of this particular design as a function of the output current.

Table 6.2 3-level converter optimized design main characteristics, referred to $I_o = 100$ mA

Inductor (L)	20.9 nH
Output capacitor (C_o)	18.6 nF
C_x capacitor (C_x)	3.8 nF
Switching frequency (f_s)	51.79 MHz
T_1 and T_3 duration	5.15 ns
T_2 duration	4.24 ns
T_1' and T_3' inactivity states duration	266 ps
Operating mode	DCM
Inductor current at the end of T_1 (I_{T_1})	202.4 mA
Total power losses (P_{losses})	46 mW
Power efficiency (η)	68.51%
Total occupied area (A_{total})	3.77 mm^2
Output voltage ripple (ΔV_o)	13.5 mV

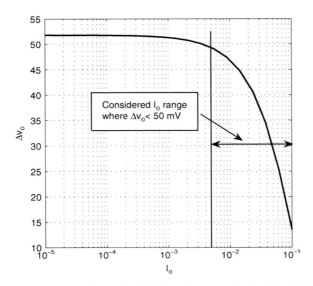

Fig. 6.15 Output voltage ripple of the optimized 3-level converter design (Fig. 6.14b), as a function of the output current

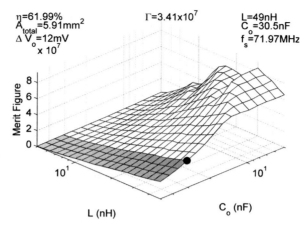

Fig. 6.16 Optimum design (that yields the maximum merit figure value) of a classical Buck converter after applying the output ripple constraint for a wide range of output current ($I_o = 5\,\text{mA} \rightarrow 100\,\text{mA}$)

For the sake of homogeneity, this results should be compared with those resulting from the classical Buck converter, applying the same output ripple constraint regarding the wide output current range. At this point, in spite of not repeating the Buck converter design space exploration with that constraint and all the corresponding results, it has been carried out again and the maximum merit figure design is shown in Fig. 6.16 ($\eta = 62\%$, $A_{total} = 5.91\,\text{mm}^2$). It is clearly observed that better performance is obtained from the 3-level converter design, since it not only increases the power efficiency by 6.5% but it reduces the occupied area in $2.14\,\text{mm}^2$.

6.3 The Selected Design to be Implemented

Although the design of Fig. 6.14b is which best balances the power efficiency and the occupied silicon area, while keeping and acceptable output ripple for a wide output current range, the design that will be finally implemented in this work is the one corresponding to Fig. 6.13f, because of its lower switching frequency. The main reason that supports this selection is that for the converter to properly work in DCM conditions, some highly specific control circuitry is required which, at the moment, the author of this work is not able to develope for such high switching frequency operation. However, it is important to lay great stress on the fact that this is not the best possible selection.

Tables from 6.3, 6.4, 6.5, 6.6, 6.7 and 6.8 present detailed data about the main converter characteristics as well as each component design.

Additionally, in Fig. 6.17 the output voltage ripple evolution as a function of I_o, when the proportional modulation of the switching frequency is applied, is exposed and contrasted against transistor-level simulation results.

Table 6.3 3-level converter selected design main characteristics, referred to $I_o = 100\,\text{mA}$

Inductor (L)	26.73 nH
Output capacitor (C_o)	25.89 nF
C_x capacitor (C_x)	5.07 nF
Switching frequency (f_s)	37.28 MHz
T_1 and T_3 duration	6.85 ns
T_2 duration	5.65 ns
T_1' and T_3' inactivity states duration	917 ps
Operating mode	DCM
Inductor current at the end of T_1 (I_{T_1})	211.2 mA
Total power losses (P_{losses})	43.5 mW
Power efficiency (η)	69.68%
Total occupied area (A_{total})	5.09 mm^2
Output voltage ripple $\rightarrow I_o = 100\,\text{mA}$ (ΔV_o)	14.6 mV
Output voltage ripple $\rightarrow I_o = 5\,\text{mA}$ (ΔV_o)	49.4 mV

Table 6.4 Inductor optimized design characteristics

Inductor design	
Number of turns (n_L)	3
External side length (s_{ext})	2.3 mm
Occupied area (A_L)	2.29 mm^2
ESR (R_L)	1.08 Ω
RMS current (I_L)	119.3 mA
Power losses (P_{L_cond})	15.4 mW

Table 6.5 C_o and C_x capacitors optimized design characteristics

Output capacitor design	
Number of cells (n)	7219
Single cells channel length (L_{ch})	3.58 μm
Single cells channel width (W)	159.5 μm
Occupied area (A_{C_o})	5.07 mm^2
ESR (R_{C_o})	17.9 mΩ
RMS current (I_{C_o})	19.3 mA
Power losses (P_{C_o})	6.7 μW
C_x capacitor design	
Occupied area (A_{C_x})	5.07 mm^2
ESR (R_{C_o})	0.1 Ω
RMS current (I_{C_o})	89.2 mA
Power losses (P_{C_o})	796.4 μW

Relative good matching appears between simulation measurements and model results although divergence occurs as the output current becomes lower. The main reasons for that might be the output capacitor ESR (not considered in the output ripple model) and a slight imbalance of the C_x capacitor voltage, due to the energy supplied to the power drivers.

In the model results, it can be observed that the output ripple is lower than 50 mV along the output current range of 5 mA $\rightarrow$ 100 mA. For output current values below

Table 6.6 3-level converter P_1 and N_1 power switches (and drivers) optimized design characteristics

P_1 (core Low-V_t transistors) and P_1-driver design	
MOSFET channel width (W_{power_MOS})	3, 930.5 μm
MOSFET channel length (L_{power_MOS})	0.3 μm
On-resistance (R_{on})	1.05 Ω
RMS current (I_{P_1})	63.1 mA
Conduction losses (P_{cond})	4.1 mW
Resistive switching losses	0.8 mW
Driver number of inverters (n)	5
Driver tapering factor (f)	4.84
Minimum inverter PMOS channel width (W_p)	1.17 μm
Minimum inverter NMOS channel width (W_n)	0.3 μm
Inverter transistors channel width (L_{ch})	0.35 μm
Driver propagation delay (t_d)	0.99 ns
Driver output fall-rise time (t_{fr})	374 ps
Driver power losses (P_{driver})	3.8 mW
N_1 (core transistors) and N_1-driver design	
MOSFET channel width (W_{power_MOS})	3, 547.2 μm
MOSFET channel length (L_{power_MOS})	0.25 μm
On-resistance (R_{on})	0.31 Ω
RMS current (I_{N_1})	101.2 mA
Conduction losses (P_{cond})	3.1 mW
Resistive switching losses	0.3 mW
Driver number of inverters (n)	5
Driver tapering factor (f)	4.58
Minimum inverter PMOS channel width (W_p)	1.46 μm
Minimum inverter NMOS channel width (W_n)	0.3 μm
Inverter transistors channel width (L_{ch})	0.35 μm
Driver propagation delay (t_d)	1 ns
Driver output fall-rise time (t_{fr})	387 ps
Driver power losses (P_{driver})	3.4 mW

5 mA, output current becomes saturated and just and slightly increase until 51.8 mV occurs at $I_o = 10\,\mu A$.

Figures 6.18 and 6.19 present the area distribution and the power losses breakdown corresponding to the selected design to be implemented.

Regarding power losses, a reduction of the inductor conduction losses is observed when compared with the classical Buck results. However, roughly, they are still a third of total losses. In respect to transistors and drivers power losses, it is observed that the equilibrium found between the PMOS and NMOS power switches for the Buck converter case is no longer kept for the four switches corresponding to the 3-level converter design. This could be caused by the use of different transistors technologies in the drivers implementation.

On the other hand, some kind of balance between overall switching and conduction losses corresponding to power switches is found: 14.7 and 12.6 mW, respectively.

Table 6.7 3-level converter P_2 and N_2 power switches (and drivers) optimized design characteristics

P_2 (core Low-V_t transistors) and P_2-driver design	
MOSFET channel width (W_{power_MOS})	6123.5 μm
MOSFET channel length (L_{power_MOS})	0.3 μm
On-resistance (R_{on})	0.66 Ω
RMS current (I_{P_2})	63.1 mA
Conduction losses (P_{cond})	2.6 mW
Resistive switching losses	0.6 mW
Driver number of inverters (n)	5
Driver tapering factor (f)	9.01
Minimum inverter PMOS channel width (W_p)	0.63 μm
Minimum inverter NMOS channel width (W_n)	0.3 μm
Inverter transistors channel width (L_{ch})	0.25 μm
Driver propagation delay (t_d)	0.79 ns
Driver output fall-rise time (t_{fr})	357 ps
Driver power losses (P_{driver})	2 mW
N_2 (core transistors) and N_2-driver design	
MOSFET channel width (W_{power_MOS})	3747.7 μm
MOSFET channel length (L_{power_MOS})	0.25 μm
On-resistance (R_{on})	0.29 Ω
RMS current (I_{N_2})	101.2 mA
Conduction losses (P_{cond})	2.8 mW
Resistive switching losses	0.4 mW
Driver number of inverters (n)	3
Driver tapering factor (f)	15.93
Minimum inverter PMOS channel width (W_p)	0.63 μm
Minimum inverter NMOS channel width (W_n)	0.3 μm
Inverter transistors channel width (L_{ch})	0.25 μm
Driver propagation delay (t_d)	0.94 ns
Driver output fall-rise time (t_{fr})	580 ps
Driver power losses (P_{driver})	1.1 mW

Table 6.8 3-level converter power switches optimized design overall characteristics

Power switches overall information	
Capacitive switching losses	2.3 mW
Total switching power losses (P_{sw})	14.8 mW
Total conduction power losses	12.5 mW
Total power losses	27.3 mW
Total occupied area	0.022 mm²

As in the classical Buck case, power losses on capacitors appear to be much lower than in the rest of the converter components. They are almost null for the output capacitor, although the become higher for the C_x capacitor due to its higher RMS current value, and its higher ESR.

Regarding area distribution, a very important imbalance is observed between the capacitors area and the inductor area. Since power efficiency is an increasing function of the inductance value when the converter is DCM operated (as exposed

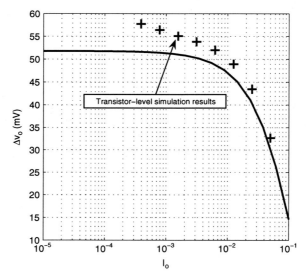

Fig. 6.17 Output voltage ripple of the implemented 3-level converter design (Fig. 6.13f), as a function of the output current

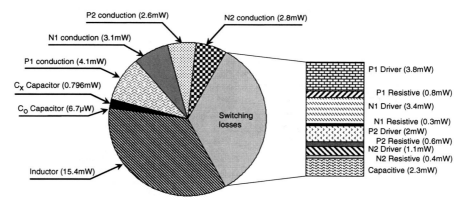

Fig. 6.18 Power losses distribution of the selected Buck converter design (for $I_o = 100\,\text{mA}$)

by Figs. 6.1, 6.2 and 6.3), it is somehow surprising that a design with a higher inductor value is not selected. The reason for this can be observed in Figs. 6.13 and 6.14, where all the optimum designs present the maximum possible inductance value (provided that all the designs yielding CCM operation are precluded).

Thus, it would be possible that a better design was found in the CCM zone, particularly, for the $I_o = 100\,\text{mA}$ condition. Nevertheless, the optimum switching frequency modulation as a function of the output current is difficult to be stated for the CCM operation. This precludes to foresee the evolution of the output ripple as a function of I_o, and even to know at which point the CCM operation would become

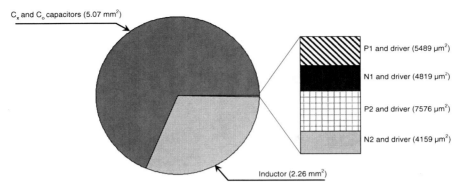

Fig. 6.19 Power losses distribution of the selected Buck converter design

DCM. That is the reason why all the CCM points were excluded from the design selection at the final stages of the 3-level converter design space exploration.

In respect to the transistors design, no apparent balanced distribution is observed. At this point, it is interesting to say that for a particular design space point (this is, a particular (L, f_s, C_o, C_x) set of design variables), several different transistors designs were obtained from the optimization by means of genetic algorithms. Since all of them presented very close values of power losses, it seems that the five-dimensional space made up by the four transistor channels lengths plus the their total power losses presents a minimum zone relatively flat (although different from each other, all the found values where in a relatively narrow range).

Chapter 7
3-Level Buck Converter Microelectronic Implementation

Abstract This chapter covers the microelectronic design and implementation of the 3-level converter design obtained in Chap. 6. The implementation of such design not only requires the integration of the reactive components and the power switches and drivers, but also includes the additional circuitry required to obtain the appropriate DCM operation.Therefore, the first section covers the design of all the required control circuits (which in this work are called as *the secondary control loop*). These include the implementation of two different control loops to assure near ideal dead-time switching, as well as the turn-off of the NMOS switches at zero inductor current. Additionally, the building blocks to generate all the required switching signals are also exposed. For all the presented circuits, the corresponding layout design is provided. In the second section, details about the power components layout are explained. Finally, the third section presents transistor-level simulation results from the complete developed system (including the power converter itself and the required control circuits), as well as the general layout distribution.

7.1 Secondary Control Loop

The microelectronic implementation of the 3-level converter design determined in Chap. 6 is based on the implementation of two different parts. The most obvious is the converter itself (inductor and capacitors, as well as power switches and drivers), whereas the second one contains all the complementary circuits that control the converter behavior to approach the expected ideal operation, particularly its switching performance. All these complementary circuits are what, in this work, is called *the secondary control loop*, to distinguish it from the conventional control loop, that acts on the duty-cycle of the control switching signal in order to get the required performance from the converter, usually output regulation or tracking. This is, the main target of the secondary control loop is to approach the converter operation to its ideal switching behavior, which is assumed by the system-level control loop designer.

In this sense, three important issues must be addressed:

- All the switching signals generation, required for the 3-level converter operation. Because of the DCM operation and the use of synchronous rectification, different

G. Villar Piqué, E. Alarcón, *CMOS Integrated Switching Power Converters*,
DOI 10.1007/978-1-4419-8843-0_7, © Springer Science+Business Media, LLC 2011

switching signals for both power PMOS switches, as well as for both NMOS switches are required separately. Moreover, an additional signal is required to identify in which half of the whole switching period the converter state is (as will be explained afterwards).

- Because of the synchronous rectification scheme and the DCM operation, NMOS switches must be turned-off when the inductor current reaches 0. Thus, specific circuitry to detect this condition is required.
- In order to reduce the switching power losses, optimum *dead-time* should be applied between the PMOS turning-off and the NMOS turning-on. In the remaining transitions, no optimum dead-time is required because the DCM operation splits this action by means of the inactivity states (T_1' and T_3').

In Fig. 7.1, the ideal waveforms of the most representative signals corresponding to the 3-level converter are depicted. The secondary control loop should be in charge of achieving that ideal switching behavior. In the figure, some important delays, required in the s and s_2 signals generation, are highlighted in gray. As explained afterwards, these delays are required for other parts of the circuitry to work.

Apart from these main functions, other interesting functionalities should be performed by the secondary control loop, such as to short-circuit the inductor during the inactivity states, to suppress inductor current and x-node voltage noisy oscillations due to the resonance between the inductor and x-node parasitic capacitances.

In Fig. 7.2, a block diagram is exposed that not only shows the building blocks corresponding to the secondary control loop, but also the signals that interact with the power plant.

As it can be observed, many signals between each block of the secondary control loop (in light gray in the figure) are required to achieve the desired functionality. Also, many signals are used to interact with the power plant. Although the details on the power plant implementation will be provided in the next section, in the following, a brief explanation list of the input and output signals of the power plant is provided:

- $\mathbf{V_o} \rightarrow$ Converter output voltage.
- $\mathbf{V_x} \rightarrow x$-node switching voltage.
- $\mathbf{ndP_1}$, $\mathbf{ndP_2} \rightarrow$ Input switching signals of the power drivers corresponding to P_1 and P_2 power switches.
- $\mathbf{dnN_1}$, $\mathbf{dnN_2} \rightarrow$ Input switching signals of the power drivers corresponding to N_1 and N_2 power switches.
- dP_1, dP_2, dN_1, $dN_2 \rightarrow$ The four power MOSFET's gate voltages.
- $\mathbf{V_A}$, $\mathbf{V_B} \rightarrow C_x$ capacitor terminals voltages.
- $\mathbf{CC_1}$, $\mathbf{CC_2} \rightarrow$ Signals used to act on 2 NMOS transistors that short-circuit the converter inductor along the inactivity states.

Only 3 external signals are required in order to drive the whole system:

- $\mathbf{CLK} \rightarrow$ An square signal used to determine the converter switching frequency. It must have double of the target switching frequency.

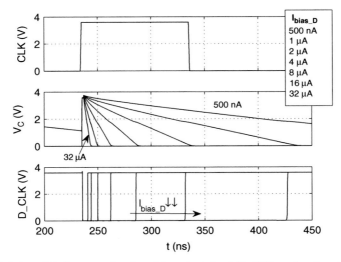

Fig. 7.5 Variation of the low-state duration of the output signal D_CLK as a function of the I_{bias_D} current (post-layout simulation results)

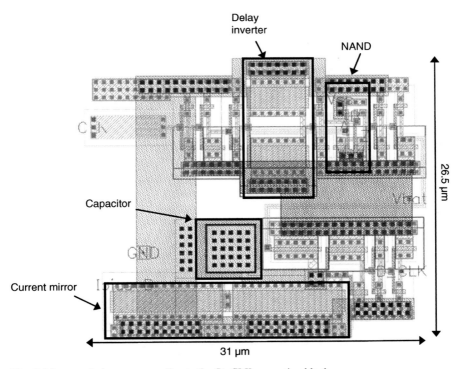

Fig. 7.6 Layout design corresponding to the D_CLK generation block

In addition to this, the circuit must generate the s signal which is a 50% duty-cycle square signal. Its frequency is the same as the switching frequency, and is used to distinguish between both halfs of the switching period. This is, its value should be high during the T_1, T_2 and T_3' states, and low during the T_3, T_2 and T_1'. Furthermore, its transitions should occur slightly later than those corresponding to the P_1 signal fall edges (as observed in Fig. 7.1), because it is required for other circuital blocks. The corresponding inverted signal s_2 is also generated in this block.

Figure 7.7 depicts the scheme corresponding to its microelectronic implementation. Regarding to the s signal generation, the circuit is based on the action of two latches ($U1$ and $U2$) that are connected in a ring formation by means of the $U3$ inverter. Since $U1$ is enabled by high-level whereas $U2$ is low-level enabled, the obtained functionality corresponds to a *Toggle* flip-flop activated by the falling-edge of the D_CLK signal. The output of the $U2$ latch is considered to be the s signal, and s_2 is obtained from the output of $U3$. The non-inverting buffers $U4$ and $U5$ are used to generate the desired delay on the s and s_2 signals, as well as to buffer them since in the layout implementation long distribution lines for these signals will be required (as it will be shown afterwards). Hence, these buffers present and even number of tapered inverters.

The generation of P_1, P_2 and their inverted signals, is based in the same idea. However, NOR functions between the corresponding outputs of the *Toggle* flip-flop (composed by $U7$, $U8$, $U9$, $U10$), and the D_CLK signal are required, since the low state duration from the later must be maintained when generating P_1 and P_2. In this case, inverters $U11$, $U12$, $U15$ and $U16$ are needed not only to invert the signals, but to buffer them.

In Fig. 7.8, post-layout simulation results are exposed. Not only the required signals generation is clearly observed, but also the required delay between the falling edges of P_1 and P_2, and the s and s_2 signals changes.

To finalize, the layout design corresponding to this circuital block is exposed in Fig. 7.9 (in this case, top metal layers routing supply lines have been removed from the image for the sake of clarity). In the figure, the most important parts of the circuit have been highlighted in black boxes.

7.1.3 dnN_1 and dnN_2 Signals Generation

This block is intended to generate the switching signals that act upon the NMOS power switches gates (by means of their corresponding gate drivers), which finally are dnN_1 and dnN_2 (see Fig. 7.2). When implementing a DCM operated synchronous rectified 3-level converter, it is necessary to turn-off the NMOS power switches when the inductor current becomes zero, in order to prevent it to reverse, which would reduce the converter output voltage, and increase both switching and conduction losses. This is due to fact that a MOS transistor is a bidirectional switch which does not turn-off automatically as its current is reversed, like diodes do. Just to sum up, note that the duration of the T_2 state must be estimated by the circuit in order to turn-off the NMOS switch when $i_L = 0$.

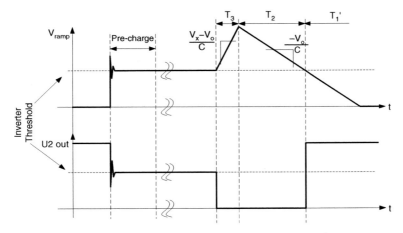

Fig. 7.11 Capacitor voltage ramp generation and voltage comparator operation

The $M5$ switch is just intended to window the current source application on the C capacitor along the T_3, T_2 and T_1' states (see Fig. 7.1).

As previously mentioned, the V_{ramp} voltage just acts as a coarse observer of the inductor current, precluding the precise turn-off of the NMOS power switches at the $i_L = 0$ moment. Inevitably, the T_2 duration estimation must be adjusted to allow such an accurate action.

As it will be explained in another section of this chapter, it is proposed to sense some converter signals so as to observe whether the NMOS power switches are being turned-off at the $i_L = 0$ condition. This information is stored in the V_{fb1} voltage (for the case of the N_1 power switch), which modifies the output current of a bidirectional current source composed by $M10$ and $M11$. As a consequence, the slope of V_{ramp} along the T_2 state is accordingly modified to adjust the right T_2 state duration. The purpose of $U1$ and $M8$ is to restrict this current adjustment to the T_2 and T_1' states.

Finally, the scheme of Fig. 7.10 shows that once the N_1 signal is generated, it is applied to the starving inverter $U8$. The purpose for that is to adjust the delay of the falling edge of the signal sent to corresponding power driver (dnN_1). This delay adjustment depends on the V_{fb_BD1} voltage, and its purpose is to obtain ideal dead-times between the P_1 turn-off and the N_1 turn-on, in the $T_1 \rightarrow T_2$ transition. It must be observed that the dnN_1 signal is inverted in respect to the N_1 signal, because of the odd number of inverters of the corresponding power driver.

Figure 7.12 exposes the post-layout simulation results of the described circuitry. Although the input signals (P_1, V_x and V_o, in the figure) have been generated by means of ideal voltage sources, the expected functionality can be clearly observed in the most significant generated signals ($U3$ out, N_1, dnN_1 and V_{ramp}). In the simulation: $V_{fb1} = 1.52$ V and $V_{fb\_BD1} = 2$ V.

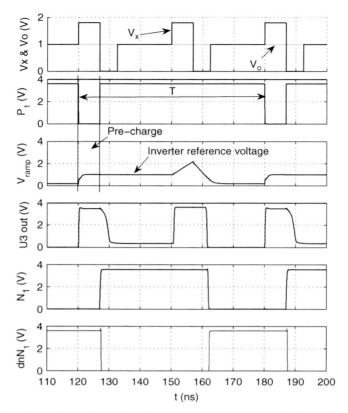

Fig. 7.12 Post-layout simulation results of the circuit that generates the dnN_1 signal

Two particularities around the voltage comparator operation are observed:

- During the pre-charge phase, the comparator reference voltage causes the output of the $U3$ inverter to go high; however, this does not affect the N_1 signal generation.
- The fast response expected from such implementation of a voltage comparator, which causes an input-to-output delay of just 945 ps.

In Fig. 7.13 the variation of the time at which N_1 signal goes low (this is, the N_1 power transistor turn-off action), can be observed as a function of the V_{fb1} voltage. In this case a sweep of several V_{fb1} values has been carried out.

To clearly show the dependence of the T_2 state duration modification as a function of the V_{fb1} voltage, it is exposed in Fig. 7.14. The results have been obtained from the transient waveforms of Fig. 7.13: T_2 duration is approximated by the time from the V_{ramp} voltage peak, until the falling edge of the $U3$ output occurs. A wide range of values (2.18 ns $\rightarrow$ 14.3 ns) with a 750 mV span is observed, and although the dependence becomes highly non-linear, it is monotonic.

The only magnitude which really identifies the presence of the shot-through event due to the simultaneous conduction of both power switches is the current spike produced by the short-circuit. Although this idea holds for a 3-level converter, it is easier to understand it in case of a classical Buck converter.

According to the ideal operation above described, no current should flow between both power transistors. Thus, just in case of a shot-through conduction a current spike flows between them. As a result, this is the magnitude to be sensed to detect this kind of event.

Unfortunately, this spike could last for tens of picosenconds, and even worse, it is not desirable to add any component to that part of the power plant, in order to avoid increasing conduction or switching losses.

Consequently, in this work it is proposed to create an observer of such situation and sense the current spike by means of a series connected capacitor. In this perspective, two smaller transistors are used to emulate the power transistors. Obviously, although their corresponding channel width is much smaller, they are of the same kind, present the same length, are supplied from the same nodes and are actuated by the same gate power drivers. In fact, in the layout design, if possible, they could be placed as one of the fingers of the power transistors, to reduce possible mismatching issues. The proposed basic idea is exposed in Fig. 7.29.

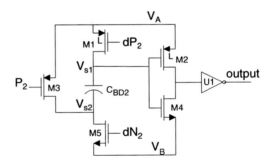

Fig. 7.29 Shot-through event detection by means of a series connected capacitor between two transistors

This circuit corresponds to the shot-through detection for the case of the P_2-N_2 transistors pair (note the voltage supply V_A-V_B from the C_x capacitor terminals), which might happen in the $T_3 \rightarrow T_2$ states transition. However, the corresponding counterpart for the P_1-N_1 pair can be straightforwardly derived.

Transistors $M1$ and $M5$ are scaled versions of both power MOSFETs, since no high power must be driven and their gates are driven by the corresponding power drivers outputs (dP_2 and dN_2), which are not expected to drive much additional capacitive load. The capacitor C_{BD2} is used to sense and storage the eventual current spikes flowing from $M1$ to $M5$. The purpose of $M3$ is to short-circuit the capacitor along the whole T_3 state (connecting both terminals to V_A), to allow the detection of a new subsequent event. $M2$ and $M4$ act as an inverter that should detect the voltage state of the V_{s1} node (this is the reason to share the same voltage supply), and the size of both transistors must be designed to properly set the corresponding threshold voltage. Finally, the $U1$ inverter should reconstruct the output of the previous inverter.

Hence, before the states transition, the V_{s1} sensing node is connected to the voltage supply V_A. After that, if no current spike is produced V_{s1} goes down to V_B. However, if the shot-through event takes place along the states transition due to simultaneous conduction, a current spike flows through both transistors and the capacitor, which stores the corresponding charge and acquires a certain voltage level depending on the strength of the shot-through event. Thus, the V_{s1} node is not equal to ground after the transition, which can be sensed by the following inverters. In other words, if the shot-through event takes place the output of $U1$ is high along the T_2 state, whereas it is low if the considered event does not happen.

According to the previous explanation, the lower the capacitor value, the higher the voltage achieved, and the easier to sense it. Unfortunately, the charge injection due to the clock-feedthrough issue across the parasitic capacitors and the gate voltage changes of $M1$ and $M5$, is also stored in the capacitor. Thus, a rather complex trade-off must be faced when scaling the power transistors, setting the capacitor value and the channel sizes of the $M2$ and $M4$ transistors.

In order to observe the C_{BD2} voltage ($V_{s1} - V_{s2}$) change when the shot-through happens, transistor-level simulation results are shown in Fig. 7.30. In this case, the dN_2 signal was delayed in respect of the dP_2, with negative ($dN2$ goes up before dP_2) and positive values. Then, the capacitor voltage is measured after the states transition (T_2 state) and depicted as function of the delay. Additionally, several curves are plotted for different capacitor values so as to observe the impact of this

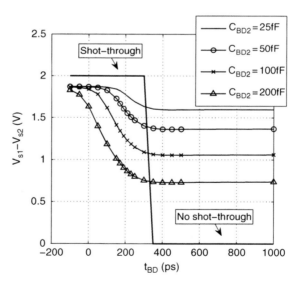

Fig. 7.30 Shot-through existence relationship with the delay between both power MOSFETs gates signals ($dP_2 \rightarrow dN_2$), and the observation of a current spike on the P_2 source terminal during the $T_3 \rightarrow T_2$ transition

magnitude upon the capacitor voltage change. Furthermore, the presence of a current spike in the P_2 source (due to the shot-through event) was observed, and it is represented as a binary magnitude, where 2 means its presence, and 0 its absence.

From these results, it is concluded that the optimum capacitor value at which the voltage change due to the current spike is maximized (which is desirable to make the detection easier), is not the lower one as expected, because of the aforementioned charge-injection issues that preclude the V_{s1} voltage to go low when no shot-through is detected. Additionally, it is observed that the capacitor is able to detect the shot-through event, storing the injected charge. In case of very low positive and even negative t_{BD} values, it is observed that the C_{BD2} capacitor voltage saturates to the supply voltage (in this case 1.8 V, as the power driver).

In the herein presented design, the capacitor values were established as $C_{BD1} = 70$ fF and $C_{BD2} = 100$ fF, as a trade-off between the capacitor voltage change and the time required to discharge it. The consequent voltage swing at the V_{s1} node must be taken into account when designing both $M2$ and $M4$ transistors, as well as the $U1$ inverter.

The results of transient transistor-level simulations are included in Fig. 7.31. Both possible cases are depicted, and the output of the $U1$ inverter along the T_2 state is observed to properly distinguish the presence or not of the shot-through conduction. To show the presence or not of the shot-through conduction, the source current of

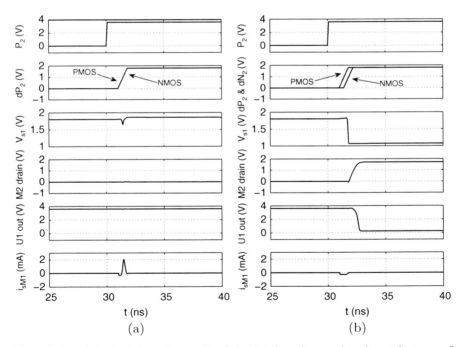

Fig. 7.31 Transistor-level simulation results of the shot-through event detection: **a** the $t_{BD} = 0$ causes the shot-through conduction; **b** $t_{BD} = 400\,ps$ precludes it

the $M1$ transistor is also shown. Clearly, the expected operation of the designed block is observed.

The circuit finally developed for the body-diode and the shot-through events correction is presented in Fig. 7.32.

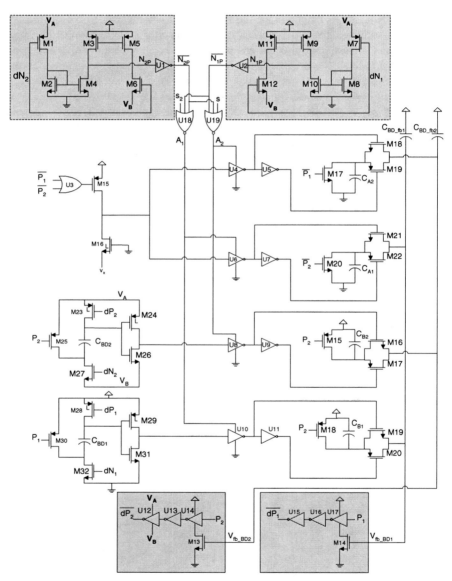

Fig. 7.32 Schematics of the microelectronic implementation corresponding to the circuital block designed to prevent the body-diode and the shot-through events

In the circuit design, the capacitor values are $C_{A1} = C_{A2} = 20\,\text{fF}$, $C_{B1} = C_{B2} = 60\,\text{fF}$. As observed, greater impact on the feedback voltage modification has been determined for the shot-through event detection. The reason for this is that it is a more dangerous event (it could damage the input battery), which requires to avoid it as soon as it appears.

The feedback capacitors C_{fb_BD1} and C_{fb_BD2} present a value of 1 pF, which is much smaller than their counterparts of the Sect. 7.1.4. This is required because four feedback loops exist in the secondary control loop, which interact between them through the converter operation. As a result, two different values were used for the feedback capacitors to separate the dynamics from these feedback loops.

Apart from the circuits that are in charge of detecting the body-diode conduction as well as the shot-through events (to modify the delay of the power MOSFETs signals generation), other required additional circuits have been highlighted by means of gray boxes.

The two boxes on the top of the figure are required to change the voltage levels of the power drivers outputs dN_1 and dN_2 from their floating voltage supply (due to the self-driving scheme, presented in Sect. 5.3), to the amplitude of the voltage supply of the control circuits (V_{bat}). These signals are required to properly determine the observation windows at which the body-diode and the shot-through events must be detected. This is carried out by means of the $U18$ and $U19$ NOR gates that supply the corresponding inverters allowing or not their operation. In Fig. 7.33, the appropriate operation of the circuit corresponding to the dN_2 signal is shown, by means of the observation of the input signal dN_2 and the outputs N_{2P} and $\overline{N_{2P}}$.

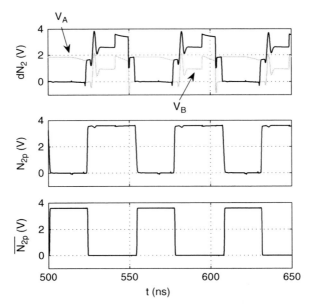

Fig. 7.33 Transistor-level simulation results of the circuit in charge of changing the voltage levels of the power driver output dN_2 (with floating power supply), to the battery voltage levels

On the bottom, two starving inverters are used to adjust the delay of the PMOS power switches driving signals, in the same way as the one used to modify the delay of the NMOS counterparts in Sect. 7.1.3.

The global operation of the whole presented circuit to adjust the dead-time when applied to the 3-level converter, can be observed in Fig. 7.34: both feedback voltages are shown, and their evolution towards the proper dead-time (t_{BD}) adjustment.

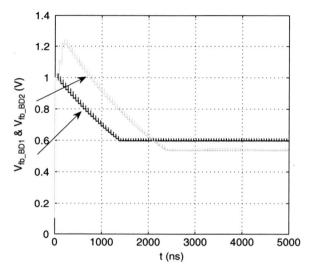

Fig. 7.34 Transistor-level simulation results from the application of the proposed circuitry to adjust the dead-time on the designed 3-level converter

The voltage glitches observed in V_{fb_BD1} and V_{fb_BD2} are due to the charge injection of the corresponding switches that connect the capacitors in parallel.

In Figs. 7.35a, b, details of the v_x voltage and the source current of both PMOS power transistors (i_{sP1} and i_{sP2}) are provided, at the beginning and at the end of the feedback loop adjustment. Power transistors gates signals are also shown. However, in this case, the source-to-gate voltage for the PMOS transistors is shown instead of the gate voltage, to better show the potential signals overlap.

In these figures, it is clear how the feedback loop is capable to adjust the dead-time until almost eliminating the body-diode conduction, without falling into the shot-through conduction. In Fig. 7.35a, two body-diode conduction events are observed that last for about 800 and 550 ps, whereas they have disappeared in Fig. 7.35b.

Additionally, no current spikes are observed in i_{sP1} and i_{sP2} in the transition towards the T_2 state. The only observed spikes occur at the beginning of T_1 and T_3 as a result of the charge spent in the parasitic capacitors of the power plant.

The developed layout design for the circuit implementation proposed in this section is presented in three different figures: Fig. 7.36 shows the main part of the

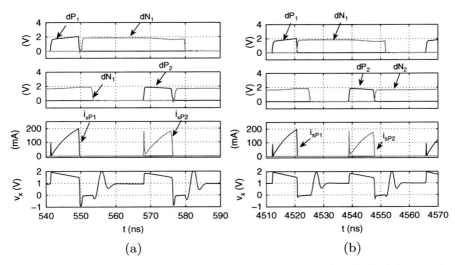

Fig. 7.35 Results of the V_{fb_BD1} and V_{fb_BD2} adjustment by means of the feedback loop, on the v_x voltage and source current of both PMOS power transistors (i_{sP1} and i_{sP2}): **a** when the voltage is still to be adjusted; **b** after the feedback loop correction

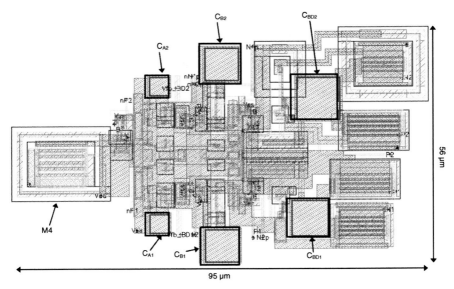

Fig. 7.36 Layout design corresponding to the main part of the circuit that adjusts the V_{fb_BD1} and V_{fb_BD2} voltages to avoid the body-diode and the shot-through conduction

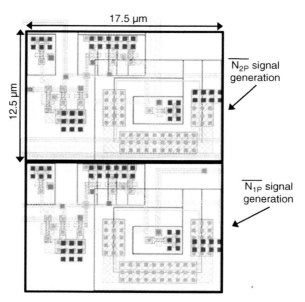

Fig. 7.37 Layout design corresponding to the circuit that changes the voltage levels of the signals applied to the gates of the NMOS power transistors

circuit, Fig. 7.37 depicts the layout design of the circuits in charge of changing the voltage levels of the NMOS power switches gates signals, and Fig. 7.38 corresponds to the layout of the circuits that modify the signals propagation delay by means of starving inverters.

Once again, the feedback capacitors have not been included in the layout view of the Fig. 7.36, because their large size might reduce the visibility of the rest of the circuit.

7.1.6 Inductor Short-Circuit

In the DCM operation of classical buck converters, noisy resonant oscillations may appear in the x-node voltage and the inductor current as well, because of the resonance between the inductor and the x-node parasitic equivalent capacitor. This kind of oscillations becomes unavoidable because of the different energy states of the capacitor at the end of T_{off} ($v_x = 0$) and at the end of T_i ($v_x = V_o$), even though the inductor current is zero. Due to the similarity between their operation, these undesired oscillations along the inactivity states also exist in a DCM operated 3-level converter. Obviously, it is desired to cancel this behavior to avoid interfering other circuits that might exist in the surroundings of the converter (specially, in case of a Powered-System-On-Chip implementation).

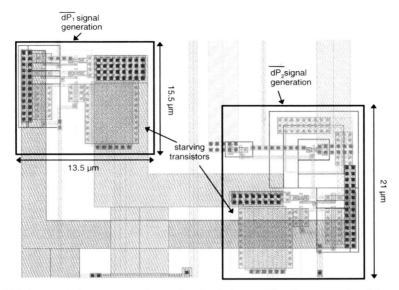

Fig. 7.38 Layout design corresponding to the circuit that modifies the propagation delay of the $\overline{dP_1}$ and $\overline{dP_2}$ signals generation

In Fig. 7.39, these oscillations can be clearly observed. They correspond to transistor-level simulation results where the selected design of the 3-level converter is implemented without the inductor short-circuit.

The most common way to suppress this issue is to short-circuit the inductor along the T_1' and T_3' states [105], to dissipate the remaining energy in the parasitic resis-

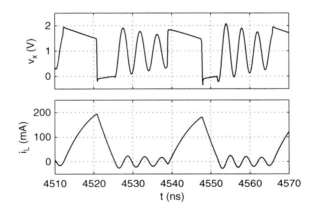

Fig. 7.39 Oscillations in v_x and i_L along the inactivity states of a 3-level converter, due to the resonance between the inductor and the parasitic capacitor of the x-node

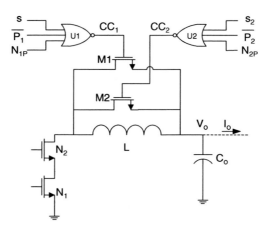

Fig. 7.40 Proposed circuit to short-circuit the inductor along the inactivity states T_1' and T_3'

tance of the short-circuit path as soon as possible. Because of the effectiveness and simplicity of this method, it has been adopted in the present design.

As exposed in Fig. 7.40, two NMOS transistors ($W_{ch} = 100\,\mu\text{m}$) are used to short-circuit the inductor. To drive these transistors, two different 3-input NOR gates determine the inactivity states, based on the information from the signals corresponding to the PMOS transistors ($\overline{P_1}$ and $\overline{P_2}$), the signals that identify the half of the switching period (s and s_2), and the gate signals of the NMOS power switches.

To guarantee that the NMOS power transistor has been switched-off before the inductor short-circuit (otherwise it could result in the output capacitor C_o short-circuit), the output of the corresponding power driver (dN_1 or dN_2) is used in the NOR gate (because it includes the signal generation and the driver delays). However, because of the floating voltage supply of the power drivers, the voltage levels of these signals need to be adapted to the control circuitry voltage supply, which is constant and equal to V_{bat}. Since this voltage adaption is already carried out in the dead-time adjustment circuit (Sect. 7.1.5), just the signals generated there are needed (N_{1P} and N_{2P}).

The results of the application of the presented circuit to the 3-level converter are shown in Fig. 7.41. In this case, v_x as well as the short-circuit signals CC_1 and CC_2 are depicted, and the effect upon the oscillations reduction is clearly observed.

In the figure, it is observed a relatively slow rising edge of both CC_1 and CC_2 signals. This is because a simple logic gate must charge and discharge the relatively large transistor that short-circuits the inductor.

To conclude this section, the designed layout corresponding to this circuit is presented in Fig. 7.42. Only the NOR gates can be observed, since $M1$ and $M2$ have been placed closer to the main inductor to avoid increasing the parasitic capacitance of the x-node.

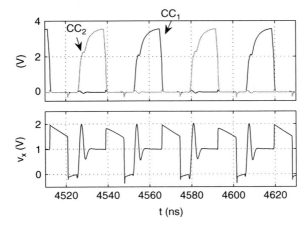

Fig. 7.41 Oscillation cancellation along the inactivity states due to the inductor short-circuit

Fig. 7.42 Layout design corresponding to the NOR gates that determine the inductor short-circuit along the converter inactivity states T_1' and T_3'

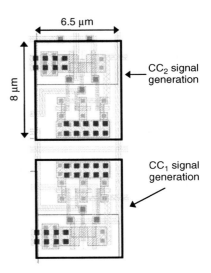

7.1.7 Overall Layout Design of the Secondary Control Loop

In Fig. 7.43 the layout design and placement of the overall secondary control loop circuitry can be observed, where the main building blocks have been highlighted by black dotted boxes. In this case, the two top metal layers have been excluded for the sake of the circuits visibility. These two layers are mainly used to implement the power supply lines, as well as to route the longest signals paths (since they present lower parasitic capacitance coupled to substrate).

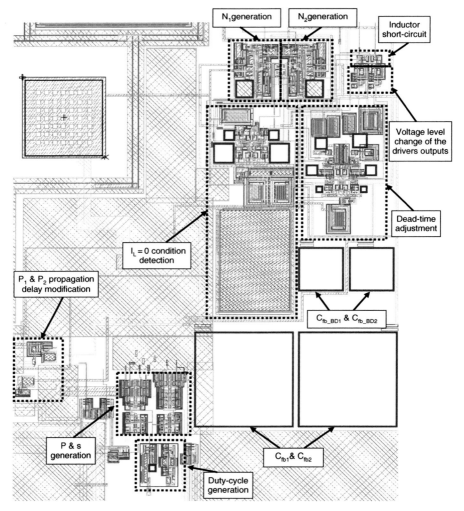

Fig. 7.43 Placement and distribution of the elements corresponding to the whole secondary control loop implementation

7.2 Power Plant Implementation

This section presents the power plant layout implementation, which includes the reactive components, the power switches and their drivers, and the two transistors used to short-circuit the inductor (as explained in Sect. 7.1.6). Since components values are determined by the design space exploration results, only the layout design is explained in each case.

7.2.1 C_o Capacitor

Recalling the design details obtained from Table 6.5, the output capacitor ($C_o =$ 25.89 nF) is composed by a matrix of 7219 MOSCAPs, each of them with a channel length and width of 3.58 and 159.5 μm, respectively. According to this data, a rather square matrix is designed, which only presents the apertures required to place the inductor bonding pads. The resulting layout distribution and its dimensions are presented in Fig. 7.44. As explained in the MOSCAP matrix presentation (Sect. 3.2), only the three bottom metal layers (*metal-1*, *-2* and *-3*) are required for its implementation.

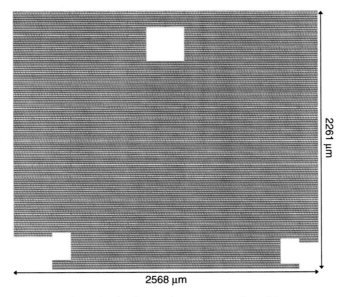

2261 μm

2568 μm

Fig. 7.44 MOSCAP matrix used to implement the output capacitor of the converter

7.2.2 C_x Capacitor

From the same table as in case of C_o, the design data corresponding to the C_x capacitor are obtained. In this case, the capacitor design should fit the same area as the occupied by the C_o capacitor (5.07 mm^2), resulting in 5.07 nF. Thus, as observed in Fig. 7.45, the C_x capacitor layout design presents a shape similar to the C_o capacitor, but in this case only the two top metal layers (*metal-4* and *-5*) are used, in addition to the special *MMC* layer that the manufacturer provides to implement MIM (*Metal-Insulator-Metal*) capacitors.

Due to technological limitations, a single large capacitor could not be designed. Instead, a matrix of 507 capacitors of 10 pF each is required.

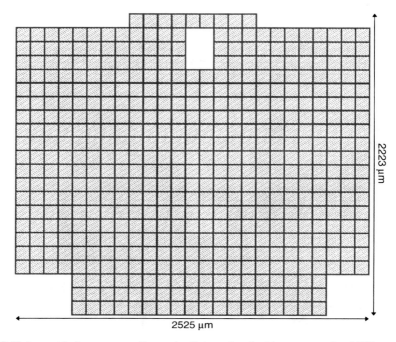

2223 μm

2525 μm

Fig. 7.45 Layout design corresponding to the C_x capacitor. In this case a matrix of 507 capacitors of 10 pF each is required

7.2.3 Inductor

From the layout design point of view, only the appropriate bonding-pads distribution is needed for the inductor implementation. According to the data from Table 6.4, a three turns spiral with an external side length of 2.3 mm is to be implemented. According to the bonding-wire manufacturer, a minimum distance between wires of 50 μm must be respected, as well as a distance between pads centers of 100 μm. Therefore, a design that accomplishes all these specifications is carried out.

Figure 7.46 shows the overall pads distribution. As expected, double pads connected by wide metal paths are used in the spiral vertexs, since two straight bonding wires need to be connected, whereas the input and output terminals of the inductor require single bonding pads. The resulting bonding wires are depicted by means of dashed gray lines.

In Fig. 7.47, more details on the designed bonding pads are shown.

7.2.4 Power Drivers Layout

The power drivers design is also based on the results of the design space exploration, presented in Chap. 6. As stated before, a particular power driver design for each different power transistor must be addressed.

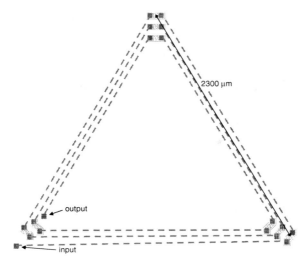

Fig. 7.46 Overall distribution of the bonding pads required to implement the triangular spiral corresponding to the inductor design

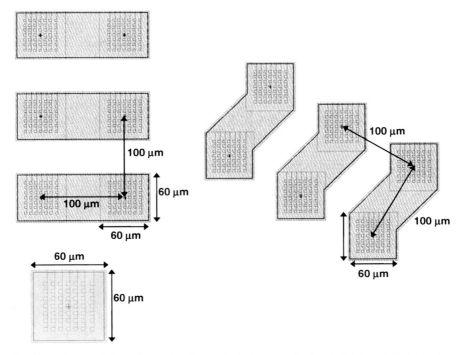

Fig. 7.47 Details of the different bonding pads designs, required to build the whole triangular spiral

Table 7.1 Channel widths of the transistors of the four power drivers (widths in microns)

Inverter		1	2	3	4	5
P_1	PMOS	1.17	5.68	27.53	133.3	646.1
	NMOS	0.3	1.45	7.04	34.11	165.24
P_2	PMOS	0.63	5.65	50.98	459.53	
	NMOS	0.3	2.7	24.38	219.76	
N_1	PMOS	1.46	6.7	30.68	140.47	643.06
	NMOS	0.3	1.37	6.29	28.78	131.78
N_2	PMOS	0.63	10	159.16		
	NMOS	0.3	4.78	76.12		

In Table 7.1, the channel widths of all the transistors that compose all the four different power drivers can be found. Since the power drivers corresponding to the P_1 and N_1 transistors make use of input – output transistors, their length is $L_{ch} = 0.34\,\mu m$. On the other hand, the core transistors that compose the drivers for the P_2 and N_2 power switches present a channel length of $L_{ch} = 0.24\,\mu m$. Additionally, in the later case, the PMOS transistors of each inverter are *low-V_T* transistor, since this selection reduces the driver energy consumption (according to the corresponding characterization developed before the drivers design).

Furthermore, all the NMOS transistors that compose the four drivers are placed in a PWELL separated from the substrate (which is possible in *Triple-Well* processes, as the one used to implement the design), in order to avoid the latch-up issue (which in this case is particularly prone to appear, because of the large size of the power drivers last stages). In case of the power drivers corresponding to the P_2 and N_2, the use of a separate PWELL for the NMOS transistors is a must, since their sources and substrates need to be connected to the floating v_B voltage, instead of ground. This is an issue that does not affect the PMOS transistors since the corresponding NWELL can always be connected to any different voltage higher than ground.

In Figs. 7.48, 7.49, 7.50 and 7.51, it is presented the layout design corresponding to the four power drivers of the P_1, P_2, N_1 and N_2 power switches, respectively.

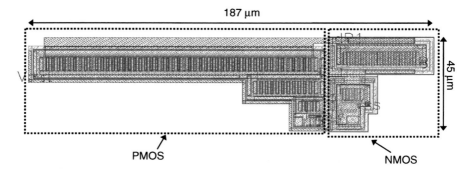

Fig. 7.48 Layout design of the power driver that acts the P_1 power transistor

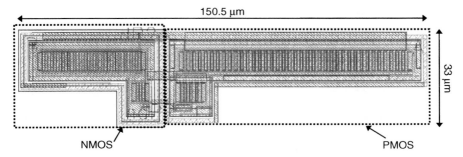

Fig. 7.49 Layout design of the power driver that acts the P_2 power transistor

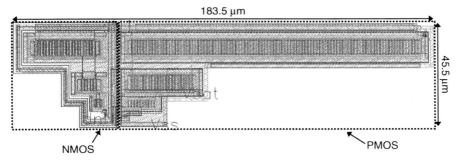

Fig. 7.50 Layout design of the power driver that acts the N_1 power transistor

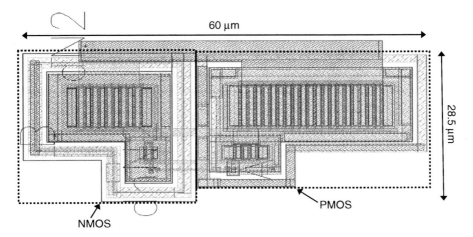

Fig. 7.51 Layout design of the power driver that acts the N_2 power transistor

The main guidelines to follow in the design of the power drivers layout is to reduce both the parasitic capacitances due to the wide metal paths interconnections, while keeping low values of the parasitic resistance of such paths. Hence, it is very interesting to put all the transistors as close as possible to each other. Although this could increase the possibilities of the latch-up issue, the use of *triple-well* NMOS transistors precludes it.

Moreover, many guard rings are placed surrounding the power drivers so as to reduce the substrate interferences on the nearby circuits, corresponding to the secondary control loop.

7.2.5 Power Transistors Layout

The developed design for each of the power MOSFETs is quite straightforward. The PMOS transistors present the *low-V_T* characteristic (since it results in lower parasitic *on*-resistance values), and NMOS are placed in a PWELL separated from the substrate to reduce the injected interferences to the substrate. Furthermore, in case of the N_2 power transistor, its placement in a separated PWELL is mandatory, since its source and body terminals are connected to v_B.

In this work, the layout design of the power MOSFETs has been determined to follow the classical finger structure so as to make it simple. Nevertheless, some guidelines have been followed in their design.

- It is interesting to use the widest possible metal paths to reduce the overall parasitic resistance of the power switch. In this perspective, it is also interesting to use as many metal layers as possible to route the power paths.
- The polysilicon gate terminals of the power transistor are connected at both ends of each finger by means of the polysilicon and the *metal-1* layers so as to reduce the gate resistance, which is directly related with the resistive switching losses (explained in Sect. 3.4.2.2).
- The wider the fingers, the higher the gate resistance because of the distributed RC circuit corresponding to the gate polysilicon. Additionally, the equivalent transistor *on*-resistance is also increased because of the narrow connections along the fingers. However, reducing too much the fingers width (W_{finger}) will increase its number (for a given transistor channel width). The later results in a width increase of the overall power transistor structure, which implies an increase of the gate connection capacitance, plus a more difficult access from the power driver output (expected to be smaller), that finally would also increase the gate resistance. Thus, the fingers length is a trade-off between the gate terminal resistance and the power switch *on-resistance*, and the gate capacitance that must be driven by the power driver.
- Furthermore, the overall size of each transistor structure is also determined by the dimensions of the terminals to be connected (this is, the inductor bonding pad, the C_x contacts, or the input and output power paths.)

- In order to further reduce the gate resistance, any power transistor is subdivided into different sections that group several fingers, and then, all of them are parallel connected. This allows to short-circuit each group gate terminals at both sides, by means of *metal-1* strips.

In Fig. 7.52 the four designed power transistors are shown. The same scale is used in all of them and their dimensions are noted, for a better comparison of their final shape and dimensions.

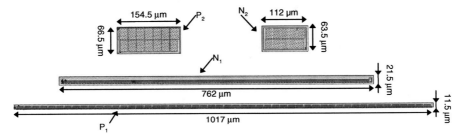

Fig. 7.52 Layout design of the four power transistors. All of them are depicted at same scale and their dimensions are noted, for a better comparison between their final shape and size

As it could be expected, P_1 and N_1 present the wider structure, since they connect the external power paths to the C_x capacitor, presenting both of them very wide contact surfaces. However, the dimensions of the corresponding drivers preclude the design of extremely wide structures (i.e. higher number of shorter fingers). On the contrary, the bonding pad used to connect the inductor presents smaller dimensions, which demands for transistor structures with lower number of longer fingers. This is the case of the transistors N_2 and P_2.

In Fig. 7.53, details of the P_1 layout design are presented as an example of the developed structures.

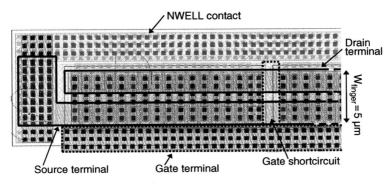

Fig. 7.53 Details of the developed layout design for the P_1 transistor

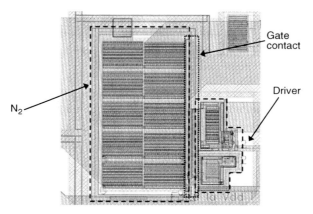

Fig. 7.54 Connection between the gate terminal of the N_2 transistor, and the output of the corresponding power driver

And finally, Fig. 7.54 shows, as an example, the connection between N_2 and its corresponding driver.

7.2.6 Overall Layout Design of the Power Plant

In this section an overall view of the whole layout design corresponding to the power plant is presented in Fig. 7.55. This figure includes all the reactive elements and the power switches and drivers.

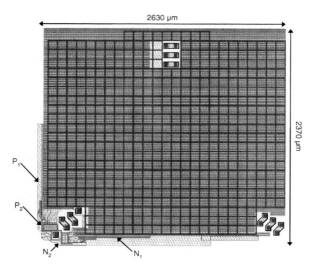

Fig. 7.55 Overall view of the layout design corresponding to the reactive elements, the power switches and drivers

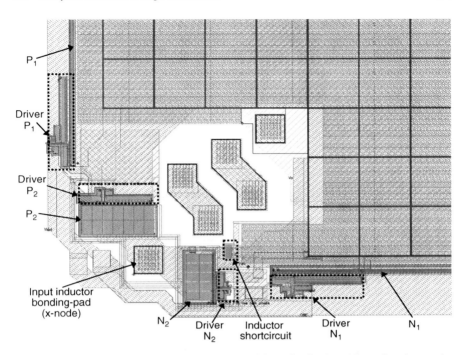

Fig. 7.56 Detailed view of the power transistors and drivers distribution, surrounding the x-node

Figure 7.56 shows a better view of the power transistors and drivers, to clarify their distribution around the bonding-pad of the inductor (x-node). Furthermore, in this figure, the transistors used to short-circuit the inductor along the inactivity states (as explained in Sect. 7.1.6) are also marked.

7.3 Complete Converter Design and Results

In this section, the most relevant results from transistor-level simulations corresponding to the complete converter design are presented. Additionally, the complete layout design joining the converter itself plus the whole secondary control loop is exposed. In all the presented waveforms, the switching frequency was stablished to $f_s = 18.64$ MHz, and the output load was set to $I_o = 50$ mA, since it was observed that the developed circuit was unable to operate at the selected switching frequency $f_s = 37.28$ MHz.

First of all, the most significant waveforms corresponding to the converter functionality are depicted. Figure 7.57 depicts the stabilization of the four feedback loops corresponding to the secondary control.

The results show that after the feedback voltages adjustment, the output voltage increases towards 1 V. Since this is achieved without any modification of the I_{bias_D}

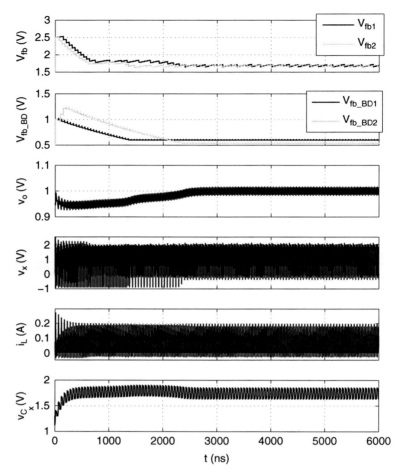

Fig. 7.57 Transistor-level simulation results of a long transient simulation of the designed 3-level converter and the secondary control loop

(that determines the T_1 and T_3 duration), it is observed that the proper dead-time adjustment and the NMOS power switches cut-off at $i_L = 0$ moment, increases the power efficiency (which results in a output voltage increase). Furthermore, an stabilization is also observed in the C_x capacitor voltage, after the feedback loops adjustment.

In the results of Fig. 7.57 it is also observed the reduction of the positive and negative peaks of the inductor current and the v_x voltage, as a result of the secondary control loop operation.

In order to provide a clearer scope on the most significant generated signals of the whole system, Fig. 7.58 depicts them for a narrow temporal span. In this case, the gate voltages of all the PMOS power transistors are shown as the source-to-gate

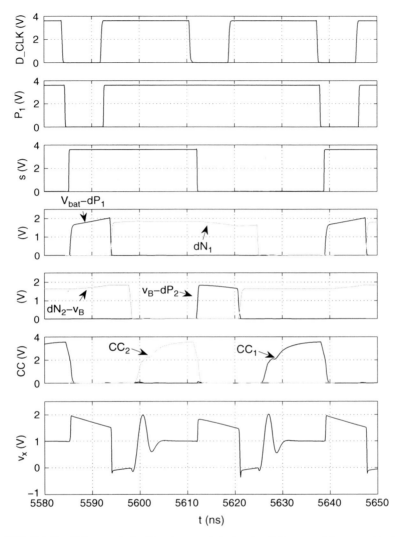

Fig. 7.58 Details of the most significant generated signals of the complete converter system (transistor-level simulations)

voltage, for a better observation of the dead-time between their falling edges and the corresponding NMOS switches gates voltages. As a visual reference, the v_x voltage is also exposed.

Figure 7.59 presents a detailed view of the most representative voltage wave-forms resulting from the operation of the complete system, such as the voltage at the terminals of the C_x capacitor (v_A and v_B) and the output voltage ripple, all of them referred to the v_x voltage waveform, so as to better identify the different converter states.

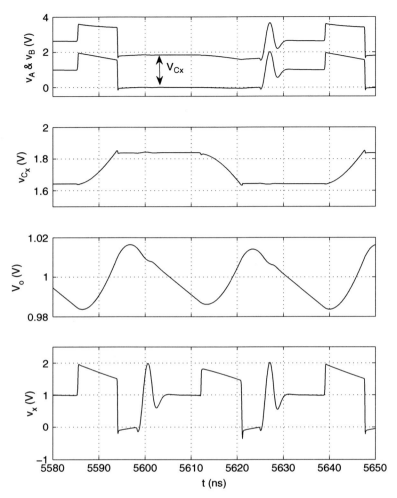

Fig. 7.59 Details of the most significant voltages signals resulting from the complete converter system operation (transistor-level simulations)

Finally, in Fig. 7.60, the most significant current waveforms are presented. In this case the input battery current (i_{bat}) is shown as well as the current sourced or provided by the C_x capacitor (i_{C_x}). In addition to this, the current sourced by the power drivers (i_{driver}) corresponding the P_1 and N_1 power switches is depicted as the difference between the battery current and the current sourced by the P_1 transistor (i_{sP1}). Inductor current is shown, as well, for which only a small oscillation is observed (thanks to the inductor short-circuit system).

From all the presented transistor-level simulation results, it is concluded that the desired operation is achieved by the proposed system, including not only the power devices, but also the secondary control loop. Unfortunately, the developed system

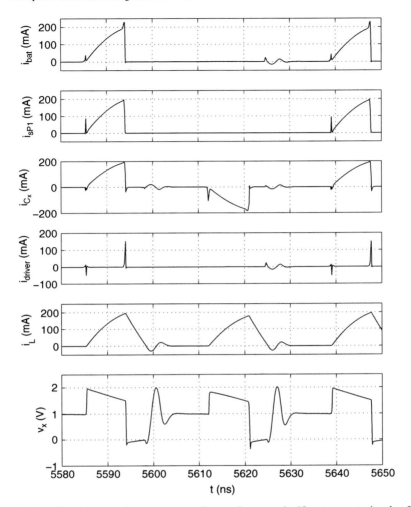

Fig. 7.60 Details of the waveforms corresponding to the most significant current signals of the converter (transistor-level simulations)

was unable the work at the obtained switching frequency from the design space exploration results carried out in Chap. 6. Consequently, the output current was reduced to 50 mA and the switching frequency to $f_s = 18.64$ MHz, in order to keep the proportionality $k = \frac{f_s}{I_o}$.

Then, the output voltage ripple (Fig. 7.62) and the power efficiency (Fig. 7.61) are measured for different output current values, while keeping the proportionality between the output current and the switching frequency (which implies to keep constant the I_{bias_D} current value).

As observed, the power efficiency obtained from the transistor-level simulation results closely matches the predicted from the design space exploration results,

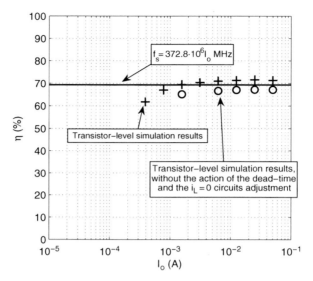

Fig. 7.61 3-level converter efficiency as a function of the output current, contrasted against the resulting from the presented models. Power efficiency corresponding to transistor-level simulation results is also depicted for the case of deactivating the adaptive dead-time and $i_L = 0$ adjustment feedback loops. In this case, adjustment voltages were set to $V_{fb1/2} = 2$ V and $V_{fb_{BD}1/2} = 0.8$ V

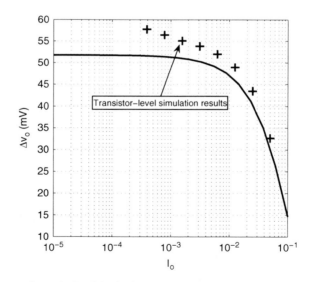

Fig. 7.62 Output voltage ripple of the implemented 3-level converter design as a function of the output current, contrasted against the resulting from the presented models

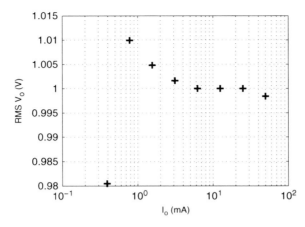

Fig. 7.63 Output voltage evolution as a function of the output current

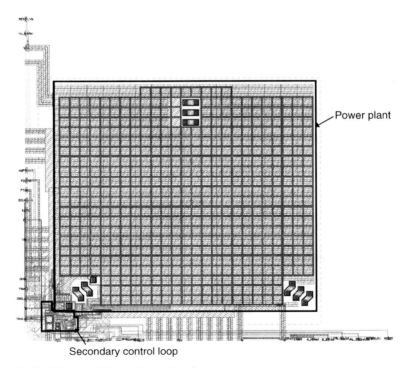

Fig. 7.64 Global view of the whole developed system

which validates the presented loss models (at least in front of the models provided by the foundry). Furthermore, it is observed that the power efficiency is kept relatively constant for about two decades of output current values. Additionally, this figure depicts transistor-level simulation results corresponding to the case of cancelling the effect of the secondary control feedback loops. In this case, the adjustment voltages were set to $V_{fb1/2} = 2\,\text{V}$ and $V_{fb_{BD}1/2} = 0.8\,\text{V}$, which produced an efficiency reduction of a 5%, approximately.

As regards the output voltage ripple, it is observed that it is higher than the level predicted from the presented models. However, this could be expected since the output ripple model presented for the 3-level converter does not include the effect of the C_o and the C_x capacitors ESRs (because of its high complexity). Moreover, it did not include the effect of the self-driving scheme to supply the power drivers from the C_x capacitor. Obviously this increases the v_{C_x} voltage swing which, in turn, increases the output ripple.

To finish with the simulation results, the evolution of the output voltage RMS value as function of the output current is exposed in Fig. 7.63. From the figure, it is observed that the output voltage is kept constant for a wide range of output current values. This implies that the power efficiency reduction as the output current becomes lower is due to the higher input current demand, rather than to the output voltage decrease.

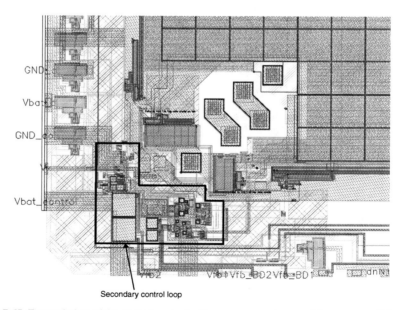

Fig. 7.65 Zoomed view of the secondary control loop components around the power switches and drivers

The layout view of the overall designed system is presented in Figs. 7.64 and 7.65 (that offers a zoomed view of the secondary control loop elements distribution around the power switches).

In the closer view of Fig. 7.65, the feedback capacitors of the blocks that adjust the T_2 state duration as well as the dead-time, can be clearly seen.

Also, other elements not presented in this chapter are observed. However, they are not part of the converter complete system, but they were added to the design just to help the test of the chip (e.g. to buffer some internal voltage signals which are needed to be observed through the package pins).

Chapter 8
Conclusions and Future Research Lines

8.1 Conclusions

Throughout the work presented in this thesis, the fully monolithic integration of switching DC-DC step-down converters has been addressed. In their integration, energy efficiency, compactness and standard CMOS implementation are pursued.

In this perspective, research has been carried out in four different areas:

- Structured design methodology both at circuit and system level, which is applied transversally to most of the designs.
- Proposals of the different converter power components IC-suitable implementation, together with their models oriented to a design space optimization.
- Topologies and converter structures more suitable to microelectronic implementation.
- Additional CMOS-compatible ancillary circuits, apart of the converter itself, that improve the converter behavior.

In this chapter, particular conclusions in all of these four areas will be presented and organized in nested lists depending on the detail level.

- **As regards to the proposed design methodology, the main conclusions derived from this thesis are:**

  - A complete design space exploration has been carried out in pursuit of an optimum design.

    - The design space exploration allows to concurrently design all the design variables of the considered circuit or system, taking into account all the interdependencies between them. The design challenge hence shifts to a modelling challenge, which has been addressed in this thesis.
    - The target microelectronic implementation of the converter offers the required access to all the design variables that allow the complete model of all the converter components, as opposite to the case of using discrete components.

G. Villar Piqué, E. Alarcón, *CMOS Integrated Switching Power Converters*,
DOI 10.1007/978-1-4419-8843-0_8, © Springer Science+Business Media, LLC 2011

- The design space exploration structure allows the existence of nested sub-explorations particularized for the design and optimization of each component.
- A proper identification and classification of the application parameters, design variables (static or dynamic), performance factors and technological information facilitates the design space exploration organization, as well as the constraints and merit figure definitions.
- In this work, the considered performance factors have been determined to be the power efficiency and the occupied area, targeting high power density.
- On the other hand, the application parameter used to constraint the design space exploration has been the output voltage ripple.

- Although this is a design methodology previously applied in other technological fields, it is a new proposal to be applied to the design of fully integrated switching power converters.
- When focusing the fully microelectronic integration of switching power converters in standard CMOS technologies, the design space exploration is a must because such an application pushes the technology limits, so that a suboptimum design could produce very poor performance.
- The transversal use of the design space exploration conforms a common conceptual framework that makes it easier the comparison and contrast between different designs that provide the same functionality (i.e. different converter topologies, components and control circuits) throughout this thesis.
- Constraints are applied on the design space exploration in terms of the allowable output voltage ripple so as to exclude those possible designs not suitable to the considered application.
- In order to obtain a single optimized design, a proper merit figure definition is required, encompassing both the energy efficiency (to be maximized) and the occupied silicon area (to be minimized).

- **With respect to the proposals of different converter components implementation and their models, the main conclusions obtained are**:

  - Models and implementation details are presented for the 4 different converter components: inductor, capacitor, power transistors and power drivers. Despite the heterogeneity of these four different components, their modelling is described as a function of the main design variables (L, f_s, C_o), which unifies the overall design procedure in the global design space exploration.
  - Regarding the inductor design and implementation:

    - The use of bonding-wire to implement on-chip inductors for power management applications in standard CMOS has been proposed and justified.
    - The development of spiral inductive structures composed by straight bonding-wire segments has been carried out.
    - It has been demonstrated that the equilateral triangular shape optimizes the inductive density while minimizes the parasitic series resistance.

- A method to design and optimize the inductor design, as a function of the number of turns and the external side length, has been proposed (similar to the design space).
- A 26.73 nH inductor with an ESR of 1.08 Ω that occupies 2.29 mm^2 has been obtained, taking advantage of the inductor optimization under the overall design space exploration. The inductor design has been corroborated by means of finite element simulation.
- The proposed bonding wire implementation allows the placement of other circuits and components underneath the inductor.

- Regarding the capacitor design and implementation:

  - It has been proposed the use of MOS capacitors (MOSCAP) to implement the output capacitor of the converter, because of its higher capacitive density.
  - The optimization of the ESR of a single MOSCAP has been proposed, from the distributed RC models of the transistor gate and channel.
  - Since the optimized ESR of a single MOSCAP has been found to be still too high for power management applications, it has been proposed a bi-dimensional matrix structure that connects a high number of single MOSCAPs to reduce the total ESR. The proposed structure makes use of the 3 bottom metal layers.
  - The ESR resulting from the whole structure has also been modelled, and a design procedure targeting an optimum design has been proposed.
  - The proposed MOSCAP matrix yields a 25.89 nF capacitor with an area occupancy of 5.07 mm^2, which compares favorably with previous state-of-the-art approaches.
  - When compared to other common on-chip capacitor implementations such as the Poly-Poly capacitors or the MIM capacitors, the proposed MOSCAP matrix appears as good trade-off between the capacitive density and the resulting ESR.

- Regarding the power drivers design and implementation:

  - A new efficiency-oriented analytical model of the tapered buffer structure has been proposed to evaluate their energy consumption as a function of the number of stages (for a given power MOSFET to be driven).
  - From the energy consumption model, the concurrent design of both the power driver and its corresponding power transistor has been proposed.

    $\triangledown$ The concepts of *intrinsic charge* and *unitary effort charge* have been proposed and used to model the current consumption of the whole tapered buffer. This model has been validated by means of transistor-level simulations.

    $\triangledown$ Additionally, a new model for the fall-rise time of the tapered buffer output has been proposed and validated by means of transistor-level simulations. This model is based on the concepts of the *intrinsic fall-rise time* and the *unitary effort fall-rise time*, that have been also introduced and explained in this work.

∇ The proposed model for the fall-rise time has been used as the link that allows the joint design of both the power MOSFET and the corresponding tapered buffer (when focusing the overall energy losses reduction).

∇ The channel width of the PMOS transistor of the unitary inverter of the tapered buffer has been proposed as one of the design variables (while keeping the NMOS channel width of the unitary inverter to the lowest possible value).

∇ Because of the high switching frequency operation of the fully integrated switching power converter, the propagation delay of the power drivers has been found to be a constraint when evaluating their design space exploration (as a function of the number of stages and the channel width of the PMOS transistor of the unitary inverter).

• Regarding the power transistors design and implementation:

  – A global model for the power transistor energy losses has been proposed.
  – From a revised existing model, the switching losses have been split into two different loss mechanisms: resistive and capacitive.

    ∇ The resistive switching losses are due to a non-instantaneous gate voltage change, which results in a no-instantaneous channel resistance change from the off-state to the on-state (while the inductor current starts to flow through the channel).

    ∇ The capacitive switching losses are due to the voltage change of all the parasitic capacitors of the power MOSFET and its circuit surroundings.

    ∇ Tables from transistor-level simulations are needed to model all the required parasitic non-linear capacitors of the power MOSFET.

    ∇ The switching power losses subdivision accounts for the loss existence even in case of Zero-Current-Switching (ZCS) or Zero-Voltage-Switching (ZVS) conditions.

  – The concurrent optimization of all the transistors of the power converter and their power drivers has been proposed.
  – The proposed model partially avoids the dependence of the switching power losses on the transistor gate resistance. The interest of this falls on the difficulty to model the gate resistance of a custom power MOSFET, since it is strongly dependent on the final layout design.
  – On the other hand, the resistive switching losses identification allows to link the transistor power losses with the fall-rise time of the power driver output.
  – Conduction losses have been modelled taking into account the variable gate voltage along the on state, in case of the 3-level converter.
  – The transistor *on*-resistance has been obtained from measurements of a complete set of simulations.
  – The concurrent optimization (towards the power losses minimization) of all power transistors (4, incase of the 3-level converter) and their

corresponding drivers, becomes a multivariable non-linear constrained problem, which could present several local minima. Unfortunately, the design space is too extended to evaluate all the possible designs.

- Consequently, *genetic algorithms* have been selected to carry out the optimization as they offer a reasonable trade-off between the global minimum search capability and the required processing time.

- In case of the C_x capacitor (for the 3-level converter), no particular proposal has been presented. Because of its floating connection, the conventional Metal-Insulator-Metal implementation (provided by the technology) has been selected, which yields a low ESR value. Interestingly, such an implementation just requires the 2 top metal layers, plus an special thin metal layer between them.
- The proposed implementation of the reactive components (L, C_o, C_x) allows their vertical placement in the layout design, which results in important occupied area savings.

■ **As regards to the research and proposals of different converter topologies more suitable to their microelectronic implementation, the main conclusions found are**:

- From typical applications, the main characteristics of the converter to be implemented have been set to: $V_{bat} = 3.6\,\text{V}$, $V_o = 1\,\text{V}$, $I_o = 100\,\text{mA}$, $\Delta V_o = 50\,\text{mV}$.
- Initially, a classical Buck converter has been considered for its fully integration on silicon, because of its simplicity.

  - The design space exploration results from the classical buck converter yielded an optimized design that offers a power efficiency of 61.02% with an area occupancy of 2.29 mm^2.
  - From the design space exploration, it has been observed that usually the highest power efficiency is obtained for DCM operation. Furthermore, when it is possible, the highest efficiency is found close to the edge between both operating modes (but in the DCM zone), since this results in a proper trade-off between switching and conduction losses.
  - The obtained power efficiency from the design space exploration has been determined to be too low. Thus, it has been suggested to investigate another step-down converter topology that presents a higher energy efficiency.
  - The observed power losses breakdown from the optimized design of the classical Buck converter have shown that about a 60% are conduction losses. As a consequence, the research for a power converter that requires a lower inductor current RMS value becomes a guideline.

- The fully integration of a 3-level two-phase converter, as a voltage regulator, has been proposed.

  - Several benefits as regards the energy losses reduction as well as the output ripple reduction are expected from the 3-level converter operation.

▽ The resulting lower inductor current RMS value yields a conduction losses reduction.

▽ From the lower voltage levels present in the x-node (approximately, half of the input voltage), a decrease of the switching losses is expected.

– On the other hand, the required additional capacitor (C_x), which classically is large enough to yield a constant voltage level, and two extra power switches and drivers suggest an increase of occupied silicon area.

– Therefore a detailed analysis considering an IC-suitable low value for the C_x capacitor has been presented covering the four different cases that arise from the 2 possible operating modes (CCM and DCM), and the 2 different cases of the output voltage being lower or higher than half of the input voltage.

– The 3-level model has been focused on the output ripple evaluation as well as the inductor current RMS value evaluation, and the RMS value of the current flowing through each of the power switches.

– Additionally, the static transfer characteristic of the 3-level converter considering low C_x values has been studied.

– A comparison between the classical Buck converter and the 3-level converter has been presented in terms of the control signal-to-output voltage characteristic, the output voltage ripple, the inductor current RMS value and the inductor switching current.

▽ The comparison between both converter topologies has been carried out for different values of the C_x capacitor, to evaluate its impact on the converter functionality, and pursue its reduction as much as possible. The common component values have been set to: $L = 35$ nH, $C_o = 30$ nF, $f_s = 25$ MHz, $V_{bat} = 3.6$ V, $V_o = 1$ V, $I_o = 100$ mA.

▽ From the comparison results it has been stated that the 3-level converter provides lower output ripple, as well as lower inductor current RMS value and lower inductor switching current, even in case of $C_x = 100$ pF.

▽ Additionally, the control signal-to-output voltage characteristic of the 3-level converter (in DCM operation) has been found to be more linear than the resulting from the classical buck converter, which could be very interesting when eventually designing control system.

– At transistor-level, a novel self-driving scheme to supply the power drivers from the C_x capacitor terminals has been proposed.

▽ This scheme guarantees that all the voltage differences related to the power switches do not exceed half of the input voltage.

▽ The proposed scheme allows the use of core transistors as the power switches (in the considered technology), with a shorter channel length. This not only reduces the conduction losses due to the lower on-resistance, but also reduces the switching losses due to the consequent smaller parasitic capacitors.

▽ The power drivers implementation corresponding to some of the power switches can also be implemented by core transistors, which in turn reduces their energy consumption.

– The design space exploration has been carried out for the 3-level converter, in order to determine an optimized design to be implemented.

▽ The addition of a new design variable (C_x) to the design space might increase its complexity and processing time to unpractical values.

▽ It has been proposed and justified to keep the 3 main design variables (as in the case of the classical Buck converter), by forcing the C_x occupied area to the same value as that occupied by the output capacitor (C_o). This links the C_x value to any value set for the C_o capacitor.

▽ The overall design space exploration results yielded a 3-level converter design that occupies 1.81 mm^2 of silicon area, while providing an energy efficiency of 69.33%. This results in a notably smaller and more efficient design than that obtained from the classical Buck converter.

▽ At this point, the proportional switching frequency modulation as a function of the output current has been revisited and has been proposed to be applied to the converter, in case of a variable output current. This way, the power efficiency is theoretically kept constant along a wide output current range, provided that the converter remains DCM operated.

▽ From the presented output ripple model, the main drawback of the proportional switching frequency modulation is the output voltage ripple increase as the output current becomes lower.

▽ From the output ripple increase observation, it has been proposed to extend the output current to a wider range ($I_o = 5\,\text{mA} \rightarrow 100\,\text{mA}$), where the output voltage ripple should be kept lower than 50 mV.

▽ The wider output current range consideration has further constrained the design space, excluding the previously found design. As a consequence, a new optimized design has been obtained, offering poorer performance factors: $\eta = 68.51\%$ and $A_{total} = 3.77\,\text{mm}^2$. In this case, the design variables have been found to be: $L = 20.9\,\text{nH}$, $C_o = 18.6\,\text{nF}$, $C_x = 3.6\,\text{nF}$, $f_s = 51.79\,\text{MHz}$.

▽ Taking into account the design constraints from the additional circuitry, finally, the selected design to be implemented has been determined to operate at a lower switching frequency.

▽ The selected design to be integrated provides a power efficiency of 69.68% and an area occupancy of 5.08 mm^2. The main characteristics of the converter are: $L = 26.73\,\text{nH}$, $C_o = 25.89\,\text{nF}$, $C_x = 5.07\,\text{nF}$, $f_s = 37.28\,\text{MHz}$ (for $I_o = 100\,\text{mA}$).

▽ From the area occupancy distribution of each component it has been concluded that:

• The power switches and drivers represent an insignificant portion of the total occupied area.

• Most of the area is occupied by the C_o (and C_x) capacitors.

- The lack of a balance between the inductor area and the capacitors area stems from the fact that a larger inductor would drive the converter into CCM operation, which increases the overall losses, because of the higher switching losses (although the conduction losses would be reduced due the lower inductor current RMS value).

■ **On the design of additional control circuits required by the 3-level converter to provide improved switching behavior, the main obtained conclusions are:**

- From the use of synchronous rectification on both the classical Buck converter and the 3-level converter (particularly for DCM operation), several non-idealities in their behavior appear: the NMOS turn-off at the $i_L = 0$ condition, the body-diode and shot-through events, and the inductor resonance with the x-node parasitic capacitor. The circuits designed to correct all these undesired issues compose what in this work has been called as the *secondary control loop*.
- After the revision of the most important error sources in conventional designs (that would result in too high power losses), a new method to adjust the NMOS switch-off action has been proposed, hence achieving improved synchronous rectification.

  - Instead of high speed and accurate circuits, the proposed method takes benefit from a feedback loop to adjust the T_2 state duration, until the Zero-Current-Switching condition is achieved.
  - The proposed method is based on the v_x voltage monitoring and the identification of events which are indicative of the inductor current sense at the moment to switch-off the NMOS power transistor.
  - The events detection is stored in a capacitor voltage, which is used to adjust the T_2 duration.
  - The use of a single transistor as a very compact and fast voltage comparator has been proposed to detect the required events that result from the NMOS switch-off before or after the $i_L = 0$ condition.
  - Proper operation has been obtained from the method as applied to the designed 3-level converter, achieving NMOS transistors switch-off at $i_L = 6.6$ mA and $i_L = 8.6$ mA (from transistor-level simulations).

- The body-diode conduction and the shot-through events have been addressed in this work, in order to approach the converter behavior to its expected operation and reduce the switching energy losses.

  - The presented method, similarly to the one used to adjust the T_2 duration, is based on a feedback loop which progressively adapts the dead-time in order to avoid both undesired issues and obtain near ideal Zero-Voltage-Switchings.
  - The body-diode conduction (due to an excessive dead-time) detection by a single transistor voltage comparator has been proposed.

- It has been proposed to detect the shot-through event (due to a too short dead-time) by means of an observer where the injected charge due to this event is stored in a capacitor, whose voltage changes with the detection.
- From transistor-level simulations, the proper functionality of the proposed method when applied to the designed 3-level converter has been observed. The dead-time is progressively adapted until the body-diode is eliminated without producing the shot-through conduction.

- It has been considered to cancel out the noisy resonance oscillations between the inductor and the x-node parasitic capacitor, by the inductor short-circuit along the inactivity states.
- The complete layout design of the whole secondary control loop and the 3-level converter has been presented.
- Proper operation of the whole developed system has been obtained by transistor-level simulations (where possible, post-layout simulations, including all the parasitic components from the layout design, have been presented).
- Global transistor-level simulations of the complete system have shown the output voltage increase as the T_2 duration and the dead-time are properly adjusted, which implies the expected power efficiency increase (since the input control signal duty-cycle remained constant).
- Unfortunately, the designed secondary control loop circuits have resulted unable to operate at the selected switching frequency of $f_s = 37.28\,\mathrm{MHz}$, according to the transistor-level simulations. As a consequence, and holding the f_s/I_o proportionality of the selected design, the output current has been set to $I_o = 50\,\mathrm{mA}$ and the switching frequency to $f_s = 18.64\,\mathrm{MHz}$.
- Simulations for different output current values (ranging from $I_o = 390\,\mu\mathrm{A}$ to $I_o = 50\,\mathrm{mA}$) have been carried out to observe the output voltage, the power efficiency and the output ripple evolutions.

- The obtained power efficiency has been found to closely match that predicted by the design space exploration results, which validates the presented loss models.
- It has been observed that the proportional switching frequency modulation as a function of the output current keeps the power efficiency rather constant for about 2 decades of I_o values.
- The predicted output ripple increase as the output current is reduced (as well as the switching frequency) has been observed, from the simulation results.
- Additionally, it has been found that the output voltage is maintained along the whole considered output current range.

8.2 Future Research Lines

From the presented work, various future research lines arise, namely:

- A proportional adjustment (instead of a binary one) of the four feedback loops of the secondary control should be implemented, since this could speed up the loops settling. This means that the correction on the feedback voltage could be proportional to the magnitude and duration of the detected event. In this respect, a study of the four secondary control loops dynamics would be interesting to optimize their response time as well as to guarantee their stability.
- The power loss comparison between the use of a diode rectifier, versus the synchronous rectification scheme, should be accurately studied, and eventually revalidated. This proposal stems from the fact that the high switching frequency operation increases the switching losses not only from the NMOS power switch but also from the required driver. Consequently, a very high frequency operation could produce as much energy losses as the resulting from the diode rectifier option.

  Additionally, if DCM operation is pursued, the use of a diode rectifier would avoid the requirement of detecting the $i_L = 0$ condition to turn-off the NMOS power switch. Even more, the body-diode and the shot-through issues would not be present.
- Obviously, the design of a primary control loop should be accomplished prior to the converter application to a final powered-system-on-chip (PSOC). It is suggested to design an efficiency oriented control that includes the Pulse-Frequency-Modulation (in order to keep the power efficiency), taking into account the particular operation and characteristics of the 3-level converter.
- The operation of the presented secondary control loop design should be tested and eventually extended to the case of $V_o > \frac{V_{bat}}{2}$.
- A seamless transition secondary control loop should be designed considering the eventual transitions between the DCM and CCM operating modes, due to the application dynamic changes (V_o, I_o).
- The effect of a converter topology that provides a larger number of voltage levels (i.e. a n-level converter) on the power losses could be evaluated. This suggested future research line arises from the fact that x-node switching voltage levels closer to the output voltage reduce the inductor current RMS value which, in turn, reduces the resulting conduction losses.
- The resulting converter dynamics should be studied, since a wide band dynamics is expected from the reactive components reduction, required to integrate them. Interesting energy-saving techniques such as the *Average Voltage Scaling* (to reduce the power consumption from digital systems) and the *Envelope Elimination and Restoration* (to improve the power efficiency of RF power amplifiers) could benefit from the expected wider band operation.
- The generated interferences (due to the inductor magnetic field) should be accurately studied, as well as their impact upon the functionality of the circuits surrounding the inductor.
- From the particular layout disposition of both converter capacitors (C_x and C_o), a capacitive coupling between them is expected. This non-ideality should be studied, in order to prevent its effects on the converter performance.

- It could be interesting to study the need of a start-up circuit and eventually develop it.
- The thermal effect of the unavoidable power losses upon the circuits surrounding the power converter should be also studied.

Appendix A
Equilateral Triangular Spiral Inductor Detailed Calculations

A.1 Self-Inductance of a Triangular-Shaped Single Inductor

The self-inductance of a single triangular coil is first calculated. This process is subdivided in the following steps:

1. The magnetic flux density (**B**) due to the current flowing along one of the triangle sides is found for any point (x, y) of the plane containing the inductor.
2. The magnetic flux (Φ) flowing through the area enclosed by half of the triangle is calculated by the integration of **B** with respect x and y, taking into account the relationship $y = f(x)$ singular of the triangle oblique side.
3. Finally, the effects upon the whole triangle are combined (in fact a multiplication by 6 is needed because of symmetry of effects), and the self-inductance is obtained by means of the ratio $L = \frac{\Phi}{I}$.

A.1.1 Application of the Biot-Savart Theorem to Calculate the Magnetic Flux Density

According to the *Biot-Savart* theorem:

$$\mathbf{B} = \frac{\mu I}{4\pi} \oint_C \frac{\mathbf{dl} \times \mathbf{a_R}}{R^2} \qquad \longrightarrow \quad \mathbf{a_R} \text{ is the director vector of } \mathbf{R} \qquad (A.1)$$

And considering the notation from Fig. A.1, the following relationships can be applied:

$$\begin{cases} \mathbf{dl} = dx_f \mathbf{a_x} \\ \mathbf{a_R} = \mathbf{a_x}\left(x - x_f\right) + \mathbf{a_y} y \\ \mathbf{dl} \times \mathbf{a_R} = \mathbf{a_R}\, y\, dx_y \end{cases} \qquad (A.2)$$

G. Villar Piqué, E. Alarcón, *CMOS Integrated Switching Power Converters*,
DOI 10.1007/978-1-4419-8843-0, © Springer Science+Business Media, LLC 2011

Fig. A.1 Basic notation for
the self-inductance
calculation of a triangular coil

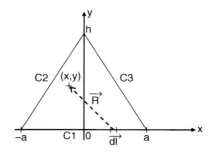

Resulting in the expression of the magnetic flux density in any (x, y) point:

$$\mathbf{B} = \mathbf{a_z} \frac{\mu I}{4\pi} \int_{-a}^{a} \frac{y \, dx_f}{\sqrt{y^2 + (x - x_f)^2}} \tag{A.3}$$

where x_f is any is the horizontal coordinate of the source point, and $\mathbf{a_z}$ is the director
vector of the magnetic flux (perpendicular to the (x, y) plane).

Here, it is interesting the following variable change: $x_f = \alpha + x \rightarrow dx_f = d\alpha$.
By means of the corresponding integration limits change, the magnetic flux density
in any point (x, y) due to the I current flowing through the side $C1$, is obtained:

$$\mathbf{B} = \mathbf{a_z} \frac{\mu I}{4\pi} \left[\frac{a + x}{y\sqrt{y^2 + (a + x)^2}} + \frac{a - x}{y\sqrt{y^2 + (a - x)^2}} \right] \tag{A.4}$$

Applying geometry the conductor thickness (radius r) can be included in the
integration limits of the triangle area to calculate the magnetic flux flowing through
it. The detail of these geometrical considerations can be found in Fig. A.2.

Therefore, the new integration limits considering the magnetic flux of half of the
triangle area are found.

$$\begin{array}{ll} x_{max} = 0 & x_{min} = \sqrt{3}r - a \\ y_{max} = \sqrt{3}a - 2r & y_{min} = r \end{array} \tag{A.5}$$

A.1.2 Magnetic Flux Density Integration Throughout Half of the Triangle Area

First, the magnetic flux density integration is carried out along the x coordinate.
Thus, integration limits are a function of y, yielding an expression equivalent to the
magnetic flux differentiation respect of y:

$$\frac{d\Phi}{dy} = \int \mathbf{B} dx \longrightarrow \begin{cases} 0 \\ \frac{y + 2r}{\sqrt{3}} - a \end{cases} \tag{A.6}$$

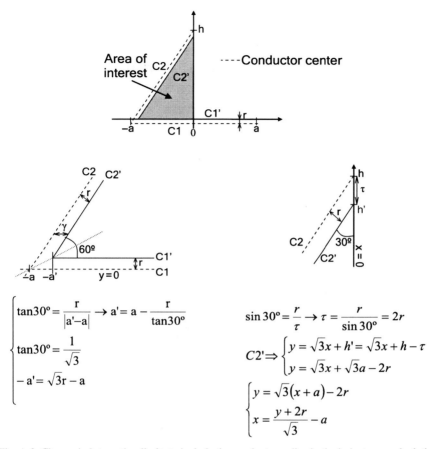

Fig. A.2 Change in integration limits to include the conductor radius in the inductance calculation

With this limits the evaluation of the first term of (A.4), by means of the variable change $k = a + x$, is as follows:

$$\frac{1}{y} \int_{\frac{y+2r}{\sqrt{3}}}^{a} \frac{k\,dk}{\sqrt{y^2 + k^2}} = \frac{\sqrt{y^2 + a^2}}{y} - \frac{2}{y\sqrt{3}} \sqrt{y^2 + r^2 + yr} \qquad (A.7)$$

And the second term of (A.4) is evaluated by means of using $k = a - x$:

$$\frac{1}{y} \int_{2a - \frac{y+2r}{\sqrt{3}}}^{a} \frac{-k\,dk}{\sqrt{y^2 + k^2}} = \frac{2}{y\sqrt{3}} \sqrt{y^2 + 3a^2 + r^2 + yr - \sqrt{3}ay - 2\sqrt{3}ar}$$

$$- \frac{\sqrt{y^2 + a^2}}{y} \qquad (A.8)$$

And merging the results from (A.7) and (A.8), the $d\Phi/dy$ expression is found:

$$\frac{d\Phi}{dy} = \frac{\mu I}{2\sqrt{3}\pi y} \left[\sqrt{y^2 + \left(r - 3\sqrt{3}a\right)y + \left(r - 3\sqrt{3}a\right)^2} - \sqrt{y^2 + yr + r^2} \right]$$

(A.9)

To integrate (A.9) respect y, it is necessary to take into account the following relationship:

$$\int \frac{\sqrt{y^2 + yb + b^2}}{y} = \sqrt{y^2 + yb + b^2} +$$

$$+ \frac{b}{2} \ln \left(\frac{b}{2} + y + \sqrt{y^2 + yb + b^2} \right) - b\,\mathrm{arctanh} \left(\frac{2b + y}{2\sqrt{y^2 + yb + b^2}} \right)$$

(A.10)

The integration of both terms of (A.9) between $\left[r, \sqrt{3}a - 2r \right]$, and the posterior sum of both results, yields the expression of the magnetic flux through one half of the triangle area, due to the current just flowing through one of the sides.

$$\Phi = \frac{\mu I}{2\sqrt{3}\pi} \left[\frac{r - \sqrt{3}a}{2} \ln \left(\frac{2k_{ar} + \sqrt{3}a - 3r}{2k_{ar} - \sqrt{3}a + 3r} \right) + \right.$$

$$+ \left(r - \sqrt{3}a\right) \left[\mathrm{arctanh} \left(\frac{\sqrt{3}a}{2k_{ar}} \right) + \mathrm{arctanh} \left(\frac{3r - 2\sqrt{3}a}{2k_{ar}} \right) \right] - k_{ar} + \sqrt{3}r -$$

$$\left. - \frac{r}{2} \ln \left(\frac{2k_{ar} + 2\sqrt{3}a - 3r}{(3 + 2\sqrt{3})r} \right) - r \left[\mathrm{arctanh} \left(\frac{\sqrt{3}}{2} \right) - \mathrm{arctanh} \left(\frac{\sqrt{3}a}{2k_{ar}} \right) \right] \right]$$

(A.11)

$$k_{ar} = \sqrt{3a^2 + 3r^2 - 3\sqrt{3}ar}$$

(A.12)

A.1.3 Self-Inductance Obtention

The final step is to multiply the magnetic flux by 6 (since the triangle has two halves and the current flows through each of the 3 sides), and dividing by the current that generates the magnetic flux.

$$L = \frac{\sqrt{3}\mu}{\pi} \left[\frac{r - \sqrt{3}a}{2} \ln \left(\frac{2k_{ar} + \sqrt{3}a - 3r}{2k_{ar} - \sqrt{3}a + 3r} \right) + \right.$$

$$+ \left(r - \sqrt{3}a \right) \left[\text{arctanh} \left(\frac{\sqrt{3}a}{2k_{ar}} \right) + \text{arctanh} \left(\frac{3r - 2\sqrt{3}a}{2k_{ar}} \right) \right] - k_{ar} + \sqrt{3}r -$$

$$\left. - \frac{r}{2} \ln \left(\frac{2k_{ar} + 2\sqrt{3}a - 3r}{(3 + 2\sqrt{3})r} \right) - r \left[\text{arctanh} \left(\frac{\sqrt{3}}{2} \right) - \text{arctanh} \left(\frac{\sqrt{3}a}{2k_{ar}} \right) \right] \right]$$

$$(A.13)$$

In Fig. A.3, the results from (A.13) are contrasted against those produced from Grover's expression [72]:

$$L = 6 \times 10^7 s_{ext} \left[\ln \left(\frac{s_{ext}}{r} \right) - 1.40546 + \frac{\mu_r}{4} \right] \qquad (A.14)$$

where μ_r is the relative permeability of the wire material, which is 1 in most cases.

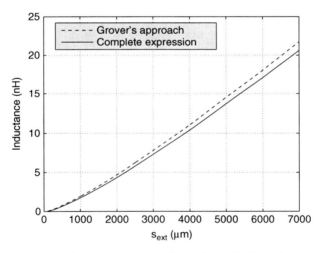

Fig. A.3 Single equilateral triangular inductance. Comparison with Grover's expression results

The comparison reveals a good matching between both expressions, although results from Grover's expression are slightly more optimistic.

A.2 Self-Inductance of the Whole Triangular Shaped Spiral

The spiral total self-inductance can be determined as the sum of the self-inductance of each coil plus the mutual inductances between all the possible combinations of pairs of coils.

$$L = \sum_{j=1}^{n_L} \sum_{i=1}^{n_L} L_{ij} \qquad\qquad (A.15)$$

In previous section the self-inductance of a triangular coil has been determined. Thus, all the terms where $i = j$ can be calculated with the aforementioned expression. On the other hand, in case of mutual inductances (L_{ij}, $i \neq j$) it is observed that $L_{ij} = L_{ji}$, by reciprocity. This means that given two different coils i and j (being i greater than j) the net magnetic flux generated by the I current of i that passes through the area of the triangle j, is equal to the net magnetic flux generated by j that passes through the area enclosed by i, provided that the same current flows through both coils. This concept is clarified n Fig. A.4.

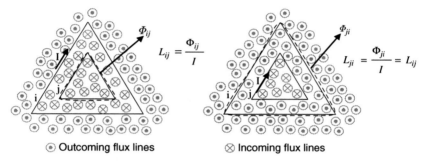

Fig. A.4 Reciprocity effect on mutual inductance of two different coils

Now, it is observed that in case that the source coil is the larger one, the computation of the mutual inductance is similar to the self-inductance expression for a single coil, but just the integration limits must be changed accordingly. Figure A.5 shows how integration limits change as the j coil becomes inner.

The new integration limits are as follows:

Integration along $x \longrightarrow \left[x = \frac{y + 2(r + p(j-i))}{\sqrt{3}} - a; \quad x = 0 \right]$

Integration along $y \longrightarrow \left[y = r + p(j-i); \quad y = \sqrt{3}a - 2(r + p(j-i)) \right]$

$$(A.16)$$

The corresponding integration of (A.4), and the subsequent product by 6 and normalization respect the I current, results in the expression of the mutual inductance L_{ij}:

$$L_{ij} = \frac{\sqrt{3}\mu}{\pi} \left[\frac{(r + p(j-i)) - \sqrt{3}a_i}{2} \ln\left(\frac{2k'_{ar} + \sqrt{3}a_i - 3(r + p(j-i))}{2k'_{ar} - \sqrt{3}a_i + 3(r + p(j-i))} \right) + \right.$$
$$\left. + \left((r + p(j-i)) - \sqrt{3}a_i \right) \right.$$

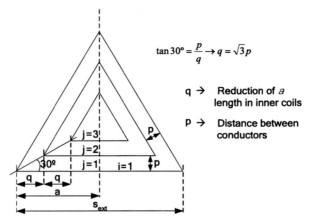

Fig. A.5 Mutul inductance calcultation. Integration limits changes as the target coil becomes inner, and far from the source

$$\left[\operatorname{arctanh}\left(\frac{\sqrt{3}a_i}{2k'_{ar}} \right) + \operatorname{arctanh}\left(\frac{3(r + p(j-i)) - 2\sqrt{3}a_i}{2k'_{ar}} \right) \right] -$$

$$-k'_{ar} + \sqrt{3}(r + p(j-i)) -$$

$$\frac{(r + p(j-i))}{2} \ln\left(\frac{2k'_{ar} + 2\sqrt{3}a_i - 3(r + p(j-i))}{(3 + 2\sqrt{3})(r + p(j-i))} \right) -$$

$$-(r + p(j-i)) \left[\operatorname{arctanh}\left(\frac{\sqrt{3}}{2} \right) - \operatorname{arctanh}\left(\frac{\sqrt{3}a_i}{2k'_{ar}} \right) \right] \right] \qquad \text{(A.17)}$$

$$k'_{ar} = \sqrt{3a_i^2 + 3(r + p(j-i))^2 - 3\sqrt{3}a_i(r + p(j-i))} \qquad \text{(A.18)}$$

$$a_i = \frac{S_{ext}}{2} - q(j-i) \qquad \text{(A.19)}$$

where a_i is the side length of the ith coil. It is noted that expression (A.17) is a more general case of the self-inductance expression (A.13), which corresponds to the particular case of $i = j$.

Finally, by means of iterative calculations where the source coil dimensions are changed (a_i), as well as the integrated area for all different coils, the total self-inductance of the triangular spiral is computed. The results for the particular case of $r = 12.5\,\mu\text{m}$ and $p = 50\,\mu\text{m}$ are shown in Fig. A.6.

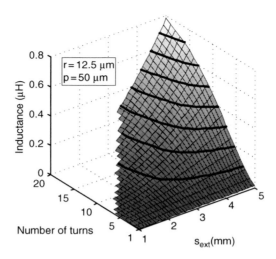

Fig. A.6 Inductance of a triangular spiral inductor as a function of the number of turns and the outer side length

From these results, some saturation of the total inductance as the number of turns increases is observed. This effect arises from the fact that as the number of turns increases while keeping constant the outer side length the inner coils become smaller, as well as their contribution to total inductance.

Appendix B
On Power Losses Related to the Capacitor Charging Process from a Constant Voltage Source

Figure B.1 shows the considered circuit: a constant voltage source is used to charge a capacitor by means of a resistive path, and a switch is used to control the charging process duration. The expression of the instantaneous current as well as the instantaneous power dissipated on the resistor due to Joule-effect are the following.

$$i(t) = \frac{V_{in}}{R} e^{\frac{-t}{RC}} \tag{B.1}$$

$$p(t) = i^2(t)R \tag{B.2}$$

Fig. B.1 Considered circuit for the lost energy evaluation purposes

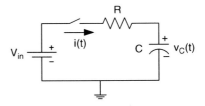

The total lost energy along the whole capacitor charging process is obtained from the instantaneous power integration between $(0, \infty)$ boundaries (considering that the initial capacitor voltage is 0).

$$E_R = \int_0^\infty i^2(t)R\,dt = \frac{V_{in}^2 C}{2} \tag{B.3}$$

On the other hand, the total energy provided by the voltage source is:

$$E_G = \int_0^\infty i(t)V_{in}(t)\,dt = V_{in}^2 C \tag{B.4}$$

Thus, it is observed that the lost energy is independent of the path resistance and that half the energy is lost after charging the capacitor from 0 to the input voltage.

This last result matches with the well-known expression of the energy stored in a capacitor $\left(\frac{1}{2}V_{in}^2 C\right)$.

Besides this straightforward result, it is very interesting to determine the energy lost in case that the capacitor charge process is stopped before the capacitor reaches the input voltage (this is, it is partially charged). In this case, the instantaneous power $p(t)$ is integrated from 0 to t.

$$E_R(t) = \int_0^t i^2(t)R\,dt = \frac{V_{in}^2 C}{2}\left[1 - e^{\frac{-2t}{RC}}\right] \tag{B.5}$$

Because the lost energy is a function of the time, as the capacitor voltage is, it can be stablished a relationship between the lost energy and the capacitor voltage:

$$E_R(t) = v_C(t)C\left[V_{in} - \frac{v_C(t)}{2}\right] \tag{B.6}$$

And the same can be stablished referred to the energy supplied by the source:

$$E_G(t) = v_C(t)CV_{in} \tag{B.7}$$

Therefore, the energy efficiency of the charging process can be expressed in terms of the final capacitor voltage (V_C).

$$\eta(\%) = \frac{E_G - E_R}{E_G} = \frac{V_C}{2V_{in}}100 \tag{B.8}$$

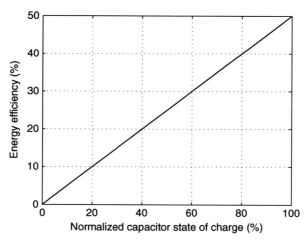

Fig. B.2 Efficiency of the capacitor charge process from a voltage source as a function of the final state of capacitor charge (normalized to the source voltage)

This results are exposed in Fig. B.2, where the final capacitor voltage is normalized to the input voltage. It is observed that the maximum attainable efficiency is just 50% (being the capacitor fully charged), whereas lower final capacitor voltage could result in much lower efficiency.

Appendix C
Proportional Switching Frequency Modulation Towards Power Efficiency Optimization for DCM Operated Converters

C.1 Switching Frequency Modulation for a Buck Converter

When addressing the implementation of a low-power converter it is very important to keep power losses as low as possible if an acceptable power efficiency is required. Often, this applications result in the DCM operation of the corresponding switching power converter, specially if wide output current range is demanded by the load, because of its own different working modes (usually intended to reduce power consumption).

In such cases, the required output current can range from few hundreds of milliamperes down to tens of microamperes. When maximum output current is provided, the power converter may operate in CCM (depending on the converter design); nevertheless, as the output current becomes lower DCM mode is unavoidable if switching frequency is kept constant.

Because of switching losses, very poor power efficiency may result when supplying low output current if constant switching frequency is maintained. Therefore, modifying properly the switching frequency as the output supplied power changes, becomes fundamental.

In [27] and [28], Arbetter *et al.*, from analytical power losses models, stated that power efficiency remains constant as the output current changes if switching frequency is set proportional to I_o, provided that the converter remains DCM operated.

$$f_s = k I_o \tag{C.1}$$

In the following an explanation on the benefits of this linear modulation of the switching frequency is provided.

The T_{on} expression for a classical Buck converter when it operates in DCM is:

$$T_{on} = \sqrt{\frac{2 I_o V_o L}{f_s V_{bat}(V_{bat} - V_o)}} \tag{C.2}$$

After applying the proportional ratio between f_s and I_o, the T_{on} duration becomes independent of the output current and the switching frequency, as well.

$$T_{on} = \sqrt{\frac{2I_o V_o L}{k I_o V_{bat}(V_{bat} - V_o)}} = \sqrt{\frac{2V_o L}{k V_{bat}(V_{bat} - V_o)}} \qquad (C.3)$$

This means that as long as the input and output voltages remain constant, the T_{on} and, in turn, T_{off} states duration are unchanged. Therefore, the inductor current waveform is exactly the same and just the inactivity state duration (T_i) is changed with the switching frequency (and the output current).

As a result, if no energy losses are considered along T_i (this is, negligible control circuitry consumption and capacitor self-discharging current), energy losses only occur in the T_{on} and T_{off} states.

Whatever energy losses are modeled (with more or less complexity), they will remain exactly the same for every inductor current pulse sent to the output capacitor. Thus, if E_{losses} is the total amount of energy losses for any time that energy is sent to the output capacitor (through T_{on} and T_{off} states), it can be easily established that total power efficiency remains constant (as stated in (C.4)).

$$\eta(\%) = \frac{P_{out}}{P_{out} + P_{losses}} \cdot 100 = \frac{V_o I_o}{V_o I_o + f_s E_{losses}} = \frac{V_o}{V_o + k E_{losses}} \qquad (C.4)$$

It is observed that after applying the (C.1) relationship, the total power efficiency becomes independent of the output current. Besides this, it is interesting to note that the proportionality between f_s and I_o establishes that no switching activity would be required in case of no supplying any current at the output.

In Fig. C.1, the resulting power efficiency as a function of output current is compared against a constant f_s Buck converter.

The results from the figure reveal the importance of reducing the frequency as the output current is reduced in order to keep an acceptable power efficiency.

Since the proportional f_s modulation is intended to keep the power efficiency as the output current changes, it becomes fundamental to get the k value that maximizes (or optimizes) the power efficiency. In case of developing analytical models for all the energy losses, an analytical procedure could be derived to find the k value that optimizes the power efficiency. However, in this work, such models are not available and the k factor is obtained by means of the design space exploration. This is, the optimum k value is determined as the ratio between the switching frequency and the considered output current for the selected design from the design space exploration results (provided that the converter is DCM operated).

Another important consequence that results from the proportional modulation of the switching frequency is that a maximum output current value exists that keeps the power converter in DCM operation (stated by (C.5) expression).

$$I_{o\_max\_DCM} = \sqrt{\frac{V_o(V_{bat} - V_o)}{2L V_{bat} k}} \qquad (C.5)$$

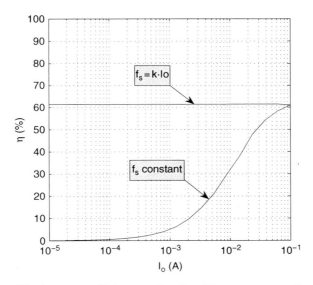

Fig. C.1 Classical Buck converter efficiency as a function of the output current. Constant switching frequency against linear modulation are compared

Beyond this upper boundary, the converter becomes CCM operated and the proportional ratio between the output current and the switching frequency, no longer guarantees that the power efficiency is optimized.

C.2 Extension to the 3-Level Converter Case

From the operation basics of the 3-level converter, all the previous reasonings could be applied to it as well.

Conceptually, the DCM operation of the 3-level Buck converter is similar to the classical Buck converter, specially is large C_x capacitor is considered. In case of low C_x value, the reasoning could be applied as well since, despite the pseudo-resonant shape of the inductor current and $v_x(t)$ waveforms, the fact that all the current pulses remain identical for any half of the switching period still holds.

This is, independently of connecting the C_x capacitor to the input voltage (T_1) or to ground (T_3), all nodes voltages and currents exhibit the same waveform. As a consequence, any power losses resulting from the converter operation will be the same for any $T_1 \rightarrow T_2$ or $T_3 \rightarrow T_2$ sequences (that are equivalent to the $T_{on} \rightarrow T_{off}$ sequence of a classical Buck converter), which leads to the same reasoning stated by (C.4).

Hence, reducing the switching frequency as the output current becomes lower will result in an increase of the T_1' and T_3' inactivity states duration.

Here, the main issue is that it is difficult to demonstrate that power efficiency remains constant as long as the switching frequency is kept proportional to the output current, because no analytical solutions were found for T_1 (and T_3) states duration. However, in Fig. C.2 the model results are exposed and compared against the case of applying a constant switching frequency. The derived benefits, in terms of power efficiency, become clear, as in case of the classical Buck. Just an overall increase of the power efficiency is observed, due to the more proper operation of the 3-level converter.

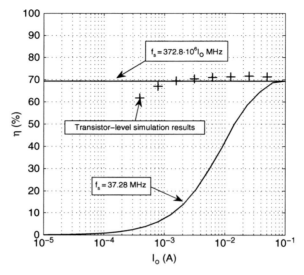

Fig. C.2 3-level converter efficiency as a function of the output current. Constant switching frequency against linear modulation are compared

In Fig. C.2, transistor-level simulation results (when $f_s = kI_o$ is applied) are depicted and contrasted against the model results. It is observed that, for more than a decade, power efficiency remains relatively constant. However, for I_o values lower than 1 mA the power efficiency is decreased, mainly due to unconsidered non-idealities such as the leakage current of the power transistors (whose effect becomes more important for such low output current).

C.3 Output Voltage Hysteretic Control as an Approach to the Linear Modulation of the Switching Frequency

C.3.1 Application to an Ideal Classical Buck Converter

In case of a classical Buck converter, one of the simplest feedback control methods is the output voltage hysteretic control [120–122]. Roughly, its operation is based

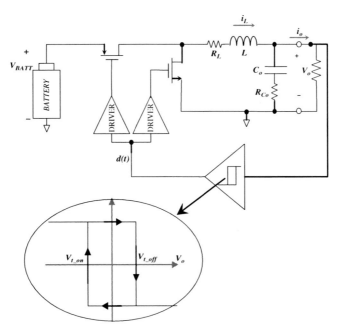

Fig. C.3 Output voltage hysteretic control of a classical Buck converter, by means of a voltage comparator

on the comparison of the output voltage against two voltage thresholds and the corresponding necessary action on the power switches (Fig. C.3).

Since the duration of the output capacitor charge and discharge processes varies with the output current value I_o, it is observed a switching frequency modulation as a function of the output load, implicit with this control method. From the parameters of an ideal buck converter and the two voltage thresholds of the hysteretic controller (V_{t_on} and V_{t_off}) an analytical expression of the resulting switching frequency can be found, by means of the corresponding expressions of the three different states of a DCM operated switching power converter (T_{on}, T_{off}, T_i).

$$T_{on} = \frac{I_o L + \sqrt{I_o^2 L^2 + 2 L C_o (V_{t\_off} - V_{t\_on})(V_{bat} - V_o)}}{V_{bat} - V_o} \tag{C.6}$$

$$T_{off} = \frac{I_o L + \sqrt{I_o^2 L^2 + 2 L C_o (V_{t\_off} - V_{t\_on})(V_{bat} - V_o)}}{V_o} \tag{C.7}$$

$$T_i = \frac{(V_{t\_off} - V_{t\_on})V_{bat}C_o}{V_o I_o} \tag{C.8}$$

$$f_s = \frac{1}{T_i + T_{on} + T_{off}} \tag{C.9}$$

From expression (C.8), it is observed that, theoretically, the hysteretic control keeps the buck converter operating in DCM for any output current value, since T_i never becomes zero (except for an infinite output current). This fact can be observed in Fig. C.4 (solid line).

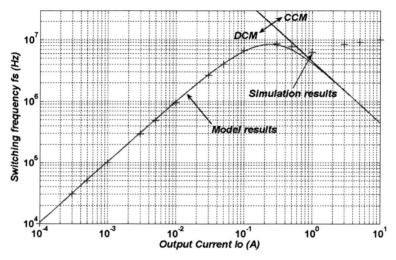

Fig. C.4 Implicit switching frequency modulation resulting from hysteretic control application contrasted with transient simulation results ($R_{C_o} = 0$)

Actually, expressions (C.6), (C.7) and (C.8) hold only as long as the low output ripple assumption holds. Thus, a hysteretic controlled buck converter does change to CCM when a certain value of output current is reached. This behavior is observed in Fig. C.4, where simulation results for f_s are depicted as a function of I_o (cross points). Additionally, it is observed that despite the asymptotical reduction of T_i towards zero (for $I_o \rightarrow \infty$), T_{on} and T_{off} are increased resulting in a switching frequency reduction that keeps the converter DCM operated.

C.3.2 Effect of the Output Capacitor ESR Upon the Switching Frequency Modulation

When the effect of the capacitor ESR (R_{C_o}) is taken into account, the behavior of the implicit frequency modulation of the hysteretic control changes and the shift from DCM to CCM appears even when the low output ripple assumption is considered.

This change is inferred from the corresponding expressions of T_{on}, T_{off} and T_i.

$$T_{on} = \frac{I_o L - R_{C_o} C_o (V_{bat} - V_o) + \sqrt{\alpha}}{V_{bat} - V_o} \tag{C.10}$$

$$T_{off} = \frac{I_o L - R_{C_o} C_o (V_{bat} - V_o) + \sqrt{\alpha}}{V_o} \tag{C.11}$$

$$T_i = \frac{V_{bat} C_o \left[R_{C_o}^2 C_o (V_{bat} - V_o) + L(V_{t\_off} - V_{t\_on}) - R_{C_o} (\sqrt{\alpha} + I_o L) \right]}{L V_o I_o} \tag{C.12}$$

$$\alpha = \left[I_o L - R_{C_o} C_o (V_{bat} - V_o) \right]^2 + 2L C_o (V_{t\_off} - V_{t\_on})(V_{bat} - V_o) \tag{C.13}$$

From expression (C.12) it can be stated that, as long as R_{C_o} is higher than zero, there is an I_o value that results in a zero T_i. Thus, if the capacitor ESR is considered, the operation mode will change from DCM to CCM. The output current value at which the operating mode changes can be derived from (C.12):

$$I_o = \frac{V_{t\_off} - V_{t\_on}}{2 R_{C_o}} \tag{C.14}$$

Again, it is confirmed that the shift from both operating modes is only possible if the R_{C_o} is higher than zero (considering the low output ripple assumption to hold). Because of the shift between both operating modes, it is necessary to find out expressions that model the switching frequency in case of CCM operation.

$$T_{on} = \frac{L(V_{t\_off} - V_{t\_on})}{R_{C_o}(V_{bat} - V_o)} \tag{C.15}$$

$$f_s = \frac{V_o}{V_{bat} T_{on}} = \frac{V_o R_{C_o}(V_{bat} - V_o)}{V_{bat} L(V_{t\_off} - V_{t\_on})} \tag{C.16}$$

It should be noted that when the Buck converter changes to CCM operation, the switching frequency becomes independent of the output current. Therefore, it will remain constant as the I_o changes. Implicit switching frequency modulation when $R_{C_o} \neq 0$ from expressions (C.10), (C.11), (C.12), (C.13), (C.14), (C.15) and (C.16) is depicted in Fig. C.5. Different values for R_{C_o} have been used, and simulation results to observe their match with theoretical expressions are also depicted (cross points), for the case of $R_{C_o} = 0.1\,\Omega$; $0.2\,\Omega$.

C.3.3 Optimum Switching Frequency Modulation vs. Output Voltage Hysteretic Control

From Figs. C.4 and C.5, it is inferred that the output voltage hysteretic control of a buck converter generates an implicit frequency modulation with the same

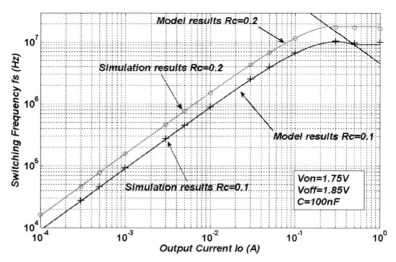

Fig. C.5 Implicit switching frequency modulation resulting from hysteretic control application contrasted with transient simulation results ($R_{C_o} = 0.1\,\Omega$ and $R_{C_o} = 0.2\,\Omega$)

trend than the optimum one, for low output current values. Since the frequency modulation resulting from the hysteretic control application as I_o changes is a curve that asymptotically approaches a linear function for low I_o values, the values of V_{t_on} and V_{t_off} can be adapted to make those trends almost coincident for a wide range of output current, resulting in optimum power efficiency for the hysteretic control application case. The relationship between V_{t_on} and V_{t_off} that produces the matching of the optimum switching frequency and the resulting one from hysteretic control, for a given output current value, is the following:

$$V_{t\_off} - V_{t\_on} = \frac{V_o}{V_{bat}kC_o} - \left[I_oL - R_{C_o}C_o(V_{bat} - V_o)\right]\sqrt{\frac{2V_o}{V_{bat}kLC_o^2(V_{bat} - V_o)}}$$

$$(C.17)$$

In Fig. C.6, the effect of the matching between both frequency modulations when (C.17) is applied can be observed. Different R_{C_o} values have been used.

Since it is observed that an increase in the capacitor ESR results in a switching frequency increase (Fig. C.6), V_{t_on} and V_{t_off} values should be adjusted for any capacitor ESR value, to improve matching with the optimum frequency modulation. It is concluded that the implicit switching frequency modulation resulting from the hysteretic control application, when it is properly designed, is very close to the optimum one for a wide range of I_o values.

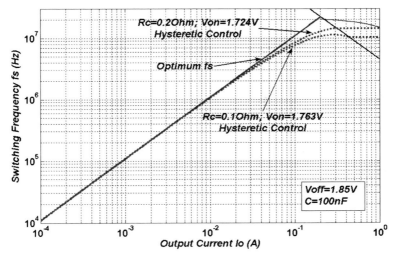

Fig. C.6 Implicit switching frequency modulation resulting from hysteretic control application with voltage thresholds adjusted to match the optimum switching frequency ($R_{C_o} = 0.1\,\Omega$ and $R_{C_o} = 0.2\,\Omega$)

C.3.4 Energy Efficiency Comparison Between Hysteretic Control and Optimized Linear Laws

Once both frequency modulations can be matched by means of the voltage thresholds modification, the consequent energy efficiency as the output current varies can be compared between the optimum linear one (presented in Sect. C.1), and that one arisen from the application of the output voltage hysteretic control.

In Fig. C.7, the obtained power efficiency for different R_{C_o} values is depicted for both frequency modulations (solid lines correspond to optimized linear laws, and dotted lines correspond to hysteretic control).

Some observations are derived from Fig. C.7:

- An increase in R_{C_o} generates lower power efficiency, due to conduction losses, for low output current values.
- The linear and quasi-linear evolutions for reduced output current values produce rather constant power efficiency at that load range, which is derived from the fact that the switching frequency is proportional to the output current.
- Then quasi-optimum linear laws are applied, as the converter approaches the CCM operation, the effect of R_{C_o} becomes insignificant.
- As I_o increases, the mismatch between the optimum switching frequency and that one generated by the hysteretic controller is increased as well. This results in the consequent lower power efficiency in case of hysteretic control.

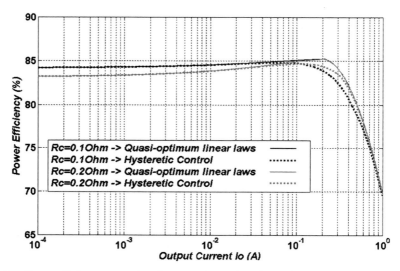

Fig. C.7 Power efficiency comparison between the quasi-optimum linear laws and the hysteretic control application ($R_{Co} = 0.1\,\Omega$ and $R_{Co} = 0.2\,\Omega$)

It is noted that an increase in R_{C_o} generates higher power efficiency at high I_o, if the hysteretic control is applied, since the implicit frequency modulation is closer to the optimum one.

C.3.5 Output Voltage Ripple Produced by the Hysteretic Control Application

Another factor to be taken into account in the complete design procedure of a switching power converter is the magnitude of its output voltage ripple. If the optimum power efficiency is targeted, increasing the switching frequency is not a possible strategy to reduce the output ripple. In case of the optimized linear laws there is no relation between the f_{s_opt} and the output capacitor value (disregarding its relationship with the corresponding ESR value). Thus, in this case the output capacitor can be increased if an output ripple reduction is required (at the cost of larger passive components that increase the converter occupied area).

When the output voltage hysteretic control is considered, a combination of voltage thresholds (V_{t_on} and V_{t_off}) and the capacitor value (C_o) is needed to concurrently keep the switching frequency modulation as close as possible to the optimum variation, and obtain the specified output voltage ripple.

Defining the output ripple as the difference between the maximum (V_{o_max}) and the minimum (V_{o_min}) values of the output voltage along a switching period, analytical expressions can be found for these values for both operating modes.

In case of DCM operation:

$$V_{o\_min} = V_{t\_on} - \frac{\left[R_{C_o}C_o(V_{bat} - V_o) - I_o L\right]^2}{2LC_o(V_{bat} - V_o)} \tag{C.18}$$

provided that the following expression holds:

$$I_o < \frac{R_{C_o}C_o(V_{bat} - V_o)}{L} \longrightarrow V_{o\_min} = V_{t\_on} \tag{C.19}$$

And for V_{o_max}:

$$V_{o\_max} = V_{t\_off} + \frac{\left[(I_{L\_max} - I_o)L - R_{C_o}C_oV_o\right]^2}{2LC_oV_o} \tag{C.20}$$

where I_{L_max} is the maximum inductor current in a switching period:

$$I_{L\_max} = \frac{I_oL - R_{C_o}C_o(V_{bat} - V_o) + \sqrt{\alpha}}{L} \tag{C.21}$$

and α is defined as in (C.13).

In this case, if the following I_o value is exceeded, V_{o_max} has V_{t_off} as lower bound:

$$I_o > I_{L\_max} - \frac{R_{C_o}C_oV_o}{L} \longrightarrow V_{o\_max} = V_{t\_off} \tag{C.22}$$

If the converter operates in continuous conduction mode:

$$V_{o\_min} = V_{t\_on} - \frac{(V_{bat} - V_o)\left[T_{on} - 2R_{C_o}C_o\right]^2}{8LC_o} \tag{C.23}$$

where T_{on} is defined as in (C.15).

In this case, V_{o_min} is limited to V_{t_on} if the following expression holds:

$$T_{on} < 2R_{C_o}C_o \longrightarrow V_{o\_min} = V_{t\_on} \tag{C.24}$$

The maximum output voltage in CCM operation is:

$$V_{o\_max} = V_{t\_off} - \frac{\left[(V_{bat} - V_o)T_{on} - 2V_oR_{C_o}C_o\right]^2}{8LC_oV_o} \tag{C.25}$$

Again, T_{on} is defined as in (C.15).

The V_{o_max} value is restricted by the following limitation:

$$T_{on} < \frac{2V_oR_{C_o}C_o}{V_{bat} - V_o} \longrightarrow V_{o\_max} = V_{t\_off} \tag{C.26}$$

Since the switching frequency is a constant function of I_o when the converter operates in CCM, it is noted that the output ripple exhibits the same trend, as stated in expressions (C.23), (C.24), (C.25) and (C.26).

It should be noted that previous expressions hold provided that the low ripple assumption holds.

The V_{o_min} and V_{o_max} evolution as a function of output current is depicted in Fig. C.8 for different R_{C_o} values. Solid lines correspond to expressions from (C.18), (C.19), (C.20), (C.21), (C.22), (C.23), (C.24), (C.25) and (C.26), while crosspoints represent transient simulations results. As observed in the figure, for $R_{C_o} = 0.2\,\Omega$ the output ripple becomes independent of the output current because of its relatively high value.

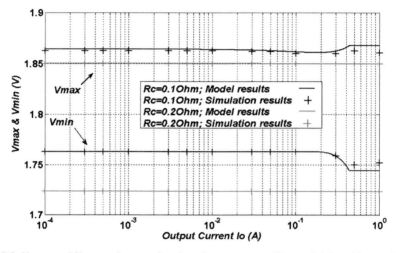

Fig. C.8 V_{o_max} and V_{o_min} values as a function of output current ($R_{Co} = 0.1\,\Omega$ and $R_{Co} = 0.2\,\Omega$)

Figure C.9 shows the output ripple ($V_{o\_max} - V_{o\_min}$ difference) as a function of the output load.

Figure C.10 shows a sample of the transient domain simulations that were carried out for the implicit frequency tuning laws resulting from the hysteretic control application. The temporal evolution of both the inductor current and the output voltage ripple are shown, for an output current linear ramp. The temporal axis is presented in a log scale in order to increase the envelope similarity with the output ripple results from Fig. C.8.

From the precedent analysis, it is observed that the output voltage ripple and the switching frequency (and in turn the resulting power efficiency), depend on the output capacitor value and the hysteresis width.

Usually, the output load current and input and output voltages are design specifications. Additionally, the inductor value can be constricted to a narrow range of values, particularly when fully integration is targeted, due to their higher difficulty to integrate them on chip.

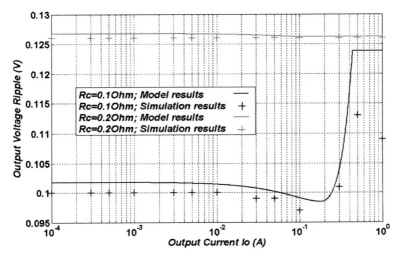

Fig. C.9 Output ripple as a function of output current ($R_{Co} = 0.1\,\Omega$ and $R_{Co} = 0.2\,\Omega$)

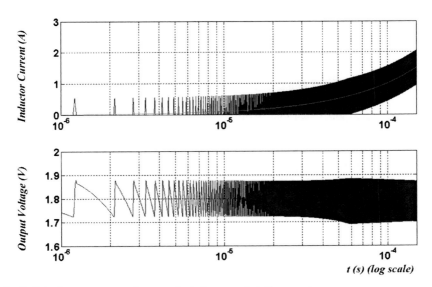

Fig. C.10 Inductor current and output ripple for a quasi-static current ramp

Hence, design guidelines can be derived from the previous analysis. In Fig. C.11, the power efficiency and the output ripple, considered as quality factors of a switching power converter, are presented as a function of the main capacitor value and the hysteresis width. Color areas correspond to pairs of capacitor-hysteresis width values that produce output ripple lower than a determined value (20, 60, 80, 100 mV, from lighter to darker, respectively). Additionally several level curves are depicted

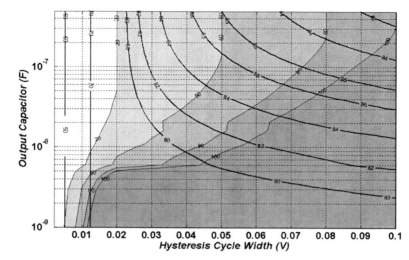

Fig. C.11 Output voltage ripple and power efficiency as a function of the output capacitor and the hysteresis cycle width

to border the areas where power efficiency is comprised in a range of values (in the figure, from left to right: 50, 70, 80, 82, 84, 85, 85, 84, 82%).

As a design example, if $L = 100$ nH ($R_L = 0.44\,\Omega$), $V_{bat} = 3.6$ V, $V_o = 1.8$ V, $I_o = 100$ mA are the design specifications, a 100 nF ($R_{C_o} = 0.1\,\Omega$) capacitor and a 60 mV hysteresis cycle, would results in a power efficiency higher than 85% and an output voltage ripple lower than 80 mV.

C.3.6 Effect of the Feedback Loop Delay Upon the Switching Frequency Modulation

Another nonideal effect to be taken into account in the description of the implicit frequency modulation arisen from the hysteretic control application is the propagation delay of its corresponding control loop. Since this delay is unavoidable in any real implementation of the control system, its effect should be considered, especially in high frequency designs such as the target application.

Although no closed analytical expressions of the corresponding frequency modulation when the propagation delay is considered could be found, its impact on the switching frequency is described in the following.

Two different delays are present in the feedback loop: propagation delays in the main transistor switch-on and switch-off transitions (t_{d_on} and t_{d_off}, respectively).

The main effect upon the switching frequency modulation is produced by t_{d_off}, since this not only increases the T_{on} state duration, but it produces a higher peak of inductor current (I_{L_max}) which results in a longer T_{off} interval. Additionally, this results in a higher voltage at the end of T_{off}, which dramatically increases the

T_i duration. In fact, the T_i duration increase is inversely proportional to the output current, resulting in a constant switching frequency reduction factor, for low output current values. This behavior can be observed in the system-level simulations in Fig. C.12, where frequency as a function of the output current is depicted for different t_{d_off} values (and a constant t_{d_on}), and compared with the ideal case ($t_{d\_on} = 0$ and $t_{d\_off} = 0$).

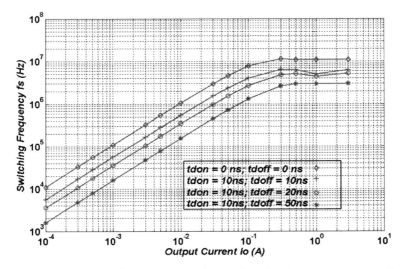

Fig. C.12 Implicit switching frequency modulation resulting from hysteretic control application, for different control loop propagation delay values (system-level simulation results)

It is observed, though counter-intuitive, that the propagation delay generates a notable reduction in switching frequency, even when the switching period is higher in several orders of magnitude. In order to avoid this, the propagation delays (t_{d_on} and t_{d_off}) should be minimized related to the T_{on} and T_{off} duration, rather than to the switching period T_s.

References

1. Wikipedia. Rechargeable battery. http://en.wikipedia.org/wiki/Rechargeable_battery#Comparison_of_battery_types.
2. MPower. Custom power solutions. http://www.mpoweruk.com.
3. UMC. http://www.umc.com.
4. International Technology Roadmap for Semiconductors. http://www.itrs.net.
5. Ban P. Wong, A. Mittal, Yu Cao, and Greg Starr. *Nano-CMOS circuit and physical design.* Wiley, New York, NY, 2005.
6. P. Larsson. Parasitic resistance in an MOS transistor used as on-chip decoupling capacitance. *IEEE Journal of Solid-State Circuits,* 32(4):574–576, April 1997.
7. Xiaodong Jin, Jia-Jiunn Ou, Chih-Hung Chen, Weidong Liu, M. J. Deen, P. R. Gray, and Chenming Hu. An effective gate resistance model for CMOS RF and noise modeling. In *Electron Devices Meeting, 1998. IEDM '98 Technical Digest., International,* pages 961–964, San Francisco, CA, December 1998.
8. M. Kazimierczuk. Collector amplitude modulation of the class e tuned power amplifier. *IEEE Transactions on Circuits and Systems,* 31(6):543–549, June 1984.
9. G. Hanington, Pin-Fan Chen, P. M. Asbeck, and L. E. Larson. High-efficiency power amplifier using dynamic power-supply voltage for CDMA applications. *IEEE Transactions on Microwave Theory and Techniques,* 47(8):1471–1476, August 1999.
10. V. Yousefzadeh, Narisi Wang, Z. Popovic, and D. Maksimovic. A digitally controlled DC/DC converter for an RF power amplifier. *IEEE Transactions on Power Electronics,* 21(1): 164–172, January 2006.
11. P. Midya, K. Haddad, L. Connell, S. Bergstedt, and B. Roeckner. Tracking power converter for supply modulation of RF power amplifiers. In *Power Electronics Specialists Conference, 2001. PESC. 2001 IEEE 32nd Annual,* volume 3, pages 1540–1545, Vancouver, BC, June 2001.
12. G. Konduri, J. Goodman, and A. Chandrakasan. Energy efficient software through dynamic voltage scheduling. In *Circuits and Systems, 1999. ISCAS '99. Proceedings of the 1999 IEEE International Symposium on,* volume 1, pages 358–361, Orlando, FL, May/June 1999.
13. M. Bhardwaj, R. Min, and A. P. Chandrakasan. Quantifying and enhancing power awareness of VLSI systems. *IEEE Transactions on Very Large Scale Integration (VLSI) Systems,* 9(6):757–772, December 2001.
14. T. Kuroda, K. Suzuki, S. Mita, T. Fujita, F. Yamane, F. Sano, A. Chiba, Y. Watanabe, K. Matsuda, T. Maeda, T. Sakurai, and T. Furuyama. Variable supply-voltage scheme for low-power high-speed CMOS digital design. In *Solid-State Circuits, IEEE Journal of,* volume 33, pages 454–462, Santa Clara, CA, March 1998.
15. M. Miyazaki, J. Kao, and A. P. Chandrakasan. A 175 mv multiply-accumulate unit using an adaptive supply voltage and body bias (ASB) architecture. In *Solid-State Circuits Conference, 2002. Digest of Technical Papers. ISSCC. 2002 IEEE International,* volume 2, pages 40–391, February 2002.

16. S. Dhar and D. Maksimovic. Switching regulator with dynamically adjustable supply voltage for low power VLSI. In *Industrial Electronics Society, 2001. IECON '01. The 27th Annual Conference of the IEEE*, volume 3, pages 1874–1879, Denver, CO, November/December 2001.

17. T. D. Burd, T. A. Pering, A. J. Stratakos, and R. W. Brodersen. A dynamic voltage scaled microprocessor system. In *Solid-State Circuits, IEEE Journal of*, volume 35, pages 1571–1580, San Francisco, CA, November 2000.

18. Dongpo Chen, Lenian He, and Xiaolang Yan. A low-dropout regulator with unconditional stability and low quiescent current. In *Communications, Circuits and Systems Proceedings, 2006 International Conference on*, volume 4, pages 2215–2218, Guilin, June 2006.

19. G. A. Rincon-Mora and P. E. Allen. A low-voltage, low quiescent current, low drop-out regulator. *IEEE Journal of Solid-State Circuits*, 33(1):36–44, January 1998.

20. C. K. Tse, S. C. Wong, and M. H. L. Chow. On lossless switched-capacitor power converters. *IEEE Transactions on Power Electronics*, 10(3):286–291, May 1995.

21. A. Ioinovici. Switched-capacitor power electronics circuits. *IEEE Circuits and Systems Magazine*, 1(3):37–42, 2001.

22. D. Maksimovic and S. Dhar. Switched-capacitor DC-DC converters for low-power on-chip applications. In *Power Electronics Specialists Conference, 1999. PESC 99. 30th Annual IEEE*, volume 1, pages 54–59, Charleston, SC, June/July 1999.

23. B. Arntzen and D. Maksimovic. Switched-capacitor DC/DC converters with resonant gate drive. *IEEE Transactions on Power Electronics*, 13(5):892–902, September 1998.

24. M. S. Makowski and D. Maksimovic. Performance limits of switched-capacitor DC-DC converters. In *Power Electronics Specialists Conference, 1995. PESC '95 Record., 26th Annual IEEE*, volume 2, pages 1215–1221, Atlanta, GA, June 1995.

25. B. Arbetter and D. Maksimovic. DC-DC converter with fast transient response and high efficiency for low-voltage microprocessor loads. In *Applied Power Electronics Conference and Exposition, 1998. APEC '98. Conference Proceedings 1998., 13th Annual*, volume 1, pages 156–162, Anaheim, CA, February 1998.

26. R. W. Erickson and D. Maksimovic. *Fundamentals of power electronics*. Kluwer Academic Publishers, Dordrecht, 2001.

27. B. Arbetter and D. Maksimovic. Control method for low-voltage DC power supply in battery-powered systems with power management. In *Power Electronics Specialists Conference, 1997. PESC '97 Record., 28th Annual IEEE*, volume 2, pages 1198–1204, St. Louis, MO, June 1997.

28. B. Arbetter, R. Erickson, and D. Maksimovic. DC-DC converter design for battery-operated systems. In *Power Electronics Specialists Conference, 1995. PESC '95 Record., 26th Annual IEEE*, volume 1, pages 103–109, Atlanta, GA, June 1995.

29. Cheung Fai Lee and P. K. T. Mok. A monolithic current-mode CMOS DC-DC converter with on-chip current-sensing technique. *IEEE Journal of Solid-State Circuits*, 39(1):3–14, January 2004.

30. I. Furukawa and Y. Sugimoto. A synchronous, step-down from 3.6 V to 1.0 V, 1 MHz PWM CMOS DC/DC converter. In *Solid-State Circuits Conference, 2001. ESSCIRC 2001. Proceedings of the 27th European*, pages 69–72, September 2001.

31. S. K. Reynolds. A DC-DC converter for short-channel CMOS technologies. *IEEE Journal of Solid-State Circuits*, 32(1):111–113, January 1997.

32. S. Sakiyama, J. Kajiwara, M. Kinoshita, K. Satomi, K. Ohtani, and A. Matsuzawa. An on-chip high-efficiency and low-noise DC/DC converter using divided switches with current control technique. In *Solid-State Circuits Conference, 1999. Digest of Technical Papers. ISSCC. 1999 IEEE International*, pages 156–157, San Francisco, CA, February 1999.

33. Won Namgoong, Mengchen Yu, and Teresa Meng. A high-efficiency variable-voltage CMOS dynamic DC-DC switching regulator. In *Solid-State Circuits Conference, 1997. Digest of Technical Papers. 44th ISSCC., 1997 IEEE International*, pages 380–381, San Francisco, CA, February 1997.

34. V. Kursun, S. G. Narendra, V. K. De, and E. G. Friedman. Analysis of buck converters for on-chip integration with a dual supply voltage microprocessor. *IEEE Transactions on Very Large Scale Integration (VLSI) Systems*, 11(3):514–522, June 2003.

35. V. Kursun, S. G. Narendra, V. K. De, and E. G. Friedman. Low-voltage-swing monolithic dc-dc conversion. *Circuits and Systems II: Express Briefs, IEEE Transactions on [see also Circuits and Systems II: Analog and Digital Signal Processing, IEEE Transactions on]*, 51(5):241–248, May 2004.

36. V. Kursun, S. G. Narendra, V. K. De, and E. G. Friedman. High input voltage step-down DC-DC converters for integration in a low voltage CMOS process. In *Quality Electronic Design, 2004. Proceedings. 5th International Symposium on*, pages 517–521, 2004.

37. S. S. Mohan, M. del Mar Hershenson, S. P. Boyd, and T. H. Lee. Simple accurate expressions for planar spiral inductances. *IEEE Journal of Solid-State Circuits*, 34(10):1419–1424, October 1999.

38. M. del Mar Hershenson, S. S. Mohan, S. P. Boyd, and T. H. Lee. Optimization of inductor circuits via geometric programming. In *Proceedings of Design Automation Conference, 1999. 36th*, pages 994–998, New Orleans, LA, June 1999.

39. C. H. Ahn and M. G. Allen. A comparison of two micromachined inductors (bar- and meander-type) for fully integrated boost DC/DC power converters. *IEEE Transactions on Power Electronics*, 11(2):239–245, March 1996.

40. C. H. Ahn and M. G. Allen. Micromachined planar inductors on silicon wafers for MEMS applications. *IEEE Transactions on Industrial Electronics*, 45(6):866–876, December 1998.

41. L. Daniel, C. R. Sullivan, and S. R. Sanders. Design of microfabricated inductors. *IEEE Transactions on Power Electronics*, 14(4):709–723, July 1999.

42. D. J. Sadler, S. Gupta, and C. H. Ahn. Micromachined spiral inductors using UV-LIGA techniques. In *Magnetics, IEEE Transactions on*, volume 37, pages 2897–2899, San Antonio, TX, July 2001.

43. S. Sugahara, M. Edo, T. Sato, and K. Yamasawa. The optimum chip size of a thin film reactor for a high-efficiency operation of a micro DC-DC converter. In *Power Electronics Specialists Conference, 1998. PESC 98 Record. 29th Annual IEEE*, volume 2, pages 1499–1503, Fukuoka, May 1998.

44. H. Nakazawa, M. Edo, Y. Katayama, M. Gekinozu, S. Sugahara, Z. Hayashi, K. Kuroki, E. Yonezawa, and K. Matsuzaki. Micro-DC/DC converter that integrates planar inductor on power IC. In *Magnetics, IEEE Transactions on*, volume 36, pages 3518–3520, Toronto, ON, September 2000.

45. Choong-Sik Kim, Seok Bae, Hee-Jun Kim, Seoung-Eui Nam, and Hyoung-June Kim. Fabrication of high frequency DC-DC converter using Ti/FeTaN film inductor. In *Magnetics, IEEE Transactions on*, volume 37, pages 2894–2896, San Antonio, TX, July 2001.

46. S. Musunuri and P. Chapman. Design of low power monolithic DC-DC buck converter with integrated inductor. *2005 IEEE 36th Conference on Power Electronics Specialists*, pages 1773–1779, September 2005.

47. E. J. Brandon, E. Wesseling, V. White, C. Ramsey, L. Del Castillo, and U. Lieneweg. Fabrication and characterization of microinductors for distributed power converters. *IEEE Transactions on Magnetics*, 39:2049–2056, July 2003.

48. S. C. O. Mathuna, T. O'Donnell, Ningning Wang, and K. Rinne. Magnetics on silicon: an enabling technology for power supply on chip. *IEEE Transactions on Power Electronics*, 20(3):585–592, May 2005.

49. S. Prabhakaran, Y. Sun, P. Dhagat, W. D. Li, and C. R. Sullivan. Microfabricated V-groove power inductors for high-current low-voltage fast-transient DC-DC converters. In *Power Electronics Specialists Conference, 2005. PESC '05. IEEE 36th*, pages 1513–1519, Recife, 2005.

50. Jun Zou, Chang Liu, D. R. Trainor, J. Chen, J. E. Schutt-Aine, and P. L. Chapman. Development of three-dimensional inductors using plastic deformation magnetic assembly (PDMA). *IEEE Transactions on Microwave Theory and Techniques*, 51:1067–1075, April 2003.

51. S. Musunuri, P. L. Chapman, Jun Zou, and Chang Liu. Design issues for monolithic DC-DC converters. *IEEE Transactions on Power Electronics*, 20(3):639–649, May 2005.
52. E. Waffenschmidt, B. Ackermann, and M. Wille. Integrated ultra thin flexible inductors for low power converters. In *Power Electronics Specialists Conference, 2005. PESC '05. IEEE 36th*, pages 1528–1534, Recife, 2005.
53. M. Ludwig, M. Duffy, T. O'Donnell, P. McCloskey, and S. C. O. Mathuna. PCB integrated inductors for low power DC/DC converter. *IEEE Transactions on Power Electronics*, 18(4):937–945, July 2003.
54. H. J. Ryu, S. D. Kim, J. J. Lee, J. Kim, S. H. Han, H. J. Kim, and C. H. Ahn. 2D and 3D simulation of toroidal type thin film inductors. In *Magnetics, IEEE Transactions on*, volume 34, pages 1360–1362, San Francisco, CA, July 1998.
55. M. del Mar Hershenson, S. S. Mohan, S. P. Boyd, and T. H. Lee. Optimization of inductor circuits via geometric programming. In *Design Automation Conference, 1999. Proceedings. 36th*, pages 994–998, New Orleans, LA, June 1999.
56. C. P. Yue and S. S. Wong. On-chip spiral inductors with patterned ground shields for SI-based RF ICs. In *Solid-State Circuits, IEEE Journal of*, volume 33, pages 743–752, Kyoto, May 1998.
57. C. P. Yue and S. S. Wong. Physical modeling of spiral inductors on silicon. *IEEE Transactions on Electron Devices*, 47(3):560–568, March 2000.
58. A. M. Niknejad and R. G. Meyer. Analysis, design, and optimization of spiral inductors and transformers for SI RF IC's. *IEEE Journal of Solid-State Circuits*, 33(10):1470–1481, October 1998.
59. Y. K. Koutsoyannopoulos and Y. Papananos. Systematic analysis and modeling of integrated inductors and transformers in RF IC design. *Circuits and Systems II: Analog and Digital Signal Processing, IEEE Transactions on [see also Circuits and Systems II: Express Briefs, IEEE Transactions on]*, 47(8):699–713, August 2000.
60. L. Vandi, P. Andreani, E. Temporiti, E. Sacchi, I. Bietti, C. Ghezzi, and R. Castello. A toroidal inductor integrated in a standard CMOS process. In *NORCHIP Conference, 2005. 23rd*, pages 39–46, January 2007.
61. N. Klemmer. Inductance calculations for MCM system design and simulation. In *Multi-Chip Module Conference, 1995. MCMC-95, Proceedings., 1995 IEEE*, pages 81–86, Santa Cruz, CA, January/February 1995.
62. Xiaoning Qi, P. Yue, T. Arnborg, H. T. Soh, H. Sakai, Zhiping Yu, and R. W. Dutton. A fast 3D modeling approach to electrical parameters extraction of bonding wires for RF circuits. *Advanced Packaging, IEEE Transactions on [see also Components, Packaging and Manufacturing Technology, Part B: Advanced Packaging, IEEE Transactions on]*, 23(3):480–488, August 2000.
63. F. Alimenti, P. Mezzanotte, L. Roselli, and R. Sorrentino. Modeling and characterization of the bonding-wire interconnection. In *Microwave Theory and Techniques, IEEE Transactions on*, volume 49, pages 142–150, Boston, MA, January 2001.
64. Chi-Taou Tsai. Package inductance characterization at high frequencies. *Components, Packaging, and Manufacturing Technology, Part B: Advanced Packaging, IEEE Transactions on [see also Components, Hybrids, and Manufacturing Technology, IEEE Transactions on]*, 17(2):175–181, May 1994.
65. Hai-Young Lee. Wideband characterization of a typical bonding wire for microwave and millimeter-wave integrated circuits. *IEEE Transactions on Microwave Theory and Techniques*, 43(1):63–68, January 1995.
66. M. Steyaert and J. Craninckx. 1.1 GHz oscillator using bondwire inductance. *Electronics Letters*, 30(3):244–245, February 1994.
67. J. Craninckx and M. Steyaert. A CMOS 1.8 GHz low-phase-noise voltage-controlled oscillator with prescaler. In *Solid-State Circuits Conference, 1995. Digest of Technical Papers. 42nd ISSCC, 1995 IEEE International*, pages 266–267, San Francisco, CA, February 1995.
68. F. Svelto and R. Castello. A bond-wire inductor-MOS varactor VCO tunable from 1.8 to 2.4 GHz. *IEEE Transactions on Microwave Theory and Techniques*, 50:403–407, January 2002.

69. M. A. L. Mostafa, J. Schlang, and S. Lazar. On-chip RF filters using bond wire inductors. In *ASIC/SOC Conference, 2001. Proceedings. 14th Annual IEEE International*, pages 98–102, Arlington, VA, September 2001.

70. K. L. R. Mertens and M. S. J. Steyaert. A 700-MHz 1-W fully differential CMOS class-E power amplifier. *IEEE Journal of Solid-State Circuits*, 37(2):137–141, February 2002.

71. A. Massarini and M. K. Kazimierczuk. Self-capacitance of inductors. *IEEE Transactions on Power Electronics*, 12(4):671–676, July 1997.

72. F. W. Grover. *Inductance calculations*. Dover Phoenix Editions, Mineola, New York, 2004.

73. D. Johns and K. Martin. *Analog integrated circuit design*. Wiley, New York, NY, 1997.

74. H. Samavati, A. Hajimiri, A. R. Shahani, G. N. Nasserbakht, and T. H. Lee. Fractal capacitors. *IEEE Journal of Solid-State Circuits*, 33(12):2035–2041, December 1998.

75. R. Aparicio and A. Hajimiri. Capacity limits and matching properties of integrated capacitors. In *Solid-State Circuits, IEEE Journal of*, volume 37, pages 384–393, San Diego, CA, March 2002.

76. S. Alenin, D. Spady, and V. Ivanov. A low ripple on-chip charge pump for bootstrapping of the noise-sensitive nodes. In *Circuits and Systems, 2006. ISCAS 2006. Proceedings. 2006 IEEE International Symposium on*, May 2006.

77. Jiandong Ge and Anh Dinh. A 0.18 μm CMOS channel select filter using Q-enhancement technique. *2004. Canadian Conference on Electrical and Computer Engineering*, 4:2143–2146, May 2004.

78. S. Chatterjee, Y. Tsividis, and P. Kinget. 0.5-V analog circuit techniques and their application in OTA and filter design. *IEEE Journal of Solid-State Circuits*, 40(12):2373–2387, December 2005.

79. Jonghoon Kim, Baekkyu Choi, Hyungsoo Kim, Woonghwan Ryu, Young hwan Yun, Seog heon Ham, Soo-Hyung Kim, Yong hee Lee, and Joungho Kim. Separated role of on-chip and on-PCB decoupling capacitors for reduction of radiated emission on printed circuit board. In *Electromagnetic Compatibility, 2001. EMC. 2001 IEEE International Symposium on*, volume 1, pages 531–536, Montreal, QC, August 2001.

80. M. D. Pant, P. Pant, and D. S. Wills. On-chip decoupling capacitor optimization using architectural level prediction. *IEEE Transactions on Very Large Scale Integration (VLSI) Systems*, 10(3):319–326, June 2002.

81. Nanju Na, T. Budell, C. Chiu, E. Tremble, and I. Wemple. The effects of on-chip and package decoupling capacitors and an efficient ASIC decoupling methodology. In *Electronic Components and Technology Conference, 2004. Proceedings. 54th*, volume 1, pages 556–567, June 2004.

82. L. Schaper, R. Ulrich, D. Nelms, E. Porter, T. Lenihan, and C. Wan. The stealth decoupling capacitor. In *Electronic Components and Technology Conference, 1997. Proceedings., 47th*, pages 724–729, San Jose, CA, May 1997.

83. D. Mannath, L. W. Schaper, and R. K. Ulrich. Advanced decoupling in high performance IC packaging. In *Electronic Components and Technology Conference, 2004. Proceedings. 54th*, volume 1, pages 266–270, June 2004.

84. S. Abedinpour and S. Kiaei. Monolithic distributed power supply for a mixed-signal integrated circuit. In *Circuits and Systems, 2003. ISCAS '03. Proceedings of the 2003 International Symposium on*, volume 3, May 2003.

85. W. H. Hayt and J. E. Kemmerly. *Análisis de circuitos en ingeniería*. McGraw Hill, New York, NY, 1993.

86. Hung Chang Lin and L. W. Linholm. An optimized output stage for MOS integrated circuits. *IEEE Journal of Solid-State Circuits*, 10(2):106–109, April 1975.

87. R. C. Jaeger and L. W. Linholm. Comments on 'an optimized output stage for MOS integrated circuits' [and reply]. *IEEE Journal of Solid-State Circuits*, 10(3):185–186, June 1975.

88. M. Nemes. Driving large capacitance in MOS LSI systems. *IEEE Journal of Solid-State Circuits*, 19(1):159–161, February 1984.

89. N. C. Li, G. L. Haviland, and A. A. Tuszynski. CMOS tapered buffer. *IEEE Journal of Solid-State Circuits*, 25(4):1005–1008, August 1990.

90. I. E. Sutherland and R. F. Sproull Logical effort: Designing for speed on the back of an envelope. In *Proceedings of the 1991 University of California/Santa Cruz Conference on Advanced Research in VLSI*, 1991.

91. Jso-Sun Choi and Kwyro Lee. Design of CMOS tapered buffer for minimum power-delay product. *IEEE Journal of Solid-State Circuits*, 29(9):1142–1145, September 1994.

92. B. S. Cherkauer and E. G. Friedman. A unified design methodology for CMOS tapered buffers. *IEEE Transactions on Very Large Scale Integration (VLSI) Systems*, 3(1):99–111, March 1995.

93. T. Sakurai and A. R. Newton. Alpha-power law MOSFET model and its applications to CMOS inverter delay and other formulas. *IEEE Journal of Solid-State Circuits*, 25(2):584–594, April 1990.

94. A. J. Stratakos, S. R. Sanders, and R. W. Brodersen. A low-voltage CMOS DC-DC converter for a portable battery-operated system. In *Power Electronics Specialists Conference, PESC '94 Record., 25th Annual IEEE*, pages 619–626, Taipei, June 1994.

95. T. Takayama and D. Maksimovic. A power stage optimization method for monolithic DC-DC converters. In *Power Electronics Specialists Conference, 2006. PESC '06. 37th IEEE*, pages 1–7, June 2006.

96. S. Musunuri and P. L. Chapman. Optimization of CMOS transistors for low power DC-DC converters. In *Power Electronics Specialists Conference, 2005. PESC '05. IEEE 36th*, pages 2151–2157, Recife, 2005.

97. T. Tolle, T. Duerbaum, and R. Elferich. Switching loss contributions of synchronous rectifiers in VRM applications. *2003. PESC '03. 2003 IEEE 34th Annual Power Electronics Specialist Conference*, volume 1, pages 144–149, June 2003.

98. Y. Bai, Y. Meng, A. Q. Huang, and F. C. Lee. A novel model for MOSFET switching loss calculation. In *Power Electronics and Motion Control Conference, 2004. IPEMC 2004. The 4th International*, volume 3, pages 1669–1672, August 2004.

99. J. Brown. Modeling the switching performance of a MOSFET in the high side of a non-isolated buck converter. *IEEE Transactions on Power Electronics*, 21(1):3–10, January 2006.

100. Yuancheng Ren, Ming Xu, Jinghai Zhou, and F. C. Lee. Analytical loss model of power MOSFET. *IEEE Transactions on Power Electronics*, 21(2):310–319, March 2006.

101. T. Lopez and R. Elferich. Method for the analysis of power MOSFET losses in a synchronous buck converter. In *12th International Power Electronics and Motion Control Conference*, pages 44–49, Portoroz, Slovenia, August 2006.

102. O. Trescases and Wai Tung Ng. Variable output, soft-switching DC/DC converter for VLSI dynamic voltage scaling power supply applications. *2004. PESC 04. 2004 IEEE 35th Annual Power Electronics Specialists Conference*, volume 6, pages 4149–4155, June 2004.

103. R. K. Williams, R. Blattner, and B. E. Mohandes. Optimization of complementary power DMOSFETs for low-voltage high-frequency DC-DC conversion. In *Applied Power Electronics Conference and Exposition, 1995. APEC '95. Conference Proceedings 1995., 10th Annual*, pages 765–772, Dallas, TX, March 1995.

104. S. Ajram and G. Salmer. Ultrahigh frequency DC-to-DC converters using gaas power switches. *IEEE Transactions on Power Electronics*, 16(5):594–602, September 2001.

105. Sang-Hwa Jung, Nam-Sung Jung, Jong-Tae Hwang, and Gyu-Hyeong Cho. An integrated CMOS DC-DC converter for battery-operated systems. In *Power Electronics Specialists Conference, 1999. PESC 99. 30th Annual IEEE*, volume 1, pages 43–47, Charleston, SC, June/July 1999.

106. O. Djekic, M. Brkovic, and A. Roy. High frequency synchronous buck converter for low voltage applications. In *Power Electronics Specialists Conference, 1998. PESC 98 Record. 29th Annual IEEE*, volume 2, pages 1248–1254, Fukuoka, May 1998.

107. P. Lopez, M. Oberst, H. Neubauer, J. Hauer, and D. Cabello. Practical considerations on doughnut transistors design. In *Circuit Theory and Design, 2005. Proceedings of the 2005 European Conference on*, volume 3, August/September 2005.

108. S. Q. Malik and R. L. Geiger. Minimization of area in low-resistance MOS switches. In *Circuits and Systems, 2000. Proceedings of the 43rd IEEE Midwest Symposium on*, volume 3, pages 1392–1395, Lansing, MI, August 2000.

109. Ming-Dou Ker, Chung-Yu Wu, and Tain-Shun Wu. Area-efficient layout design for CMOS output transistors. *IEEE Transactions on Electron Devices*, 44(4):635–645, April 1997.

110. A. Van den Bosch, M. S. J. Steyaert, and W. Sansen. A high-density, matched hexagonal transistor structure in standard CMOS technology for high-speed applications. In *Semiconductor Manufacturing, IEEE Transactions on*, volume 13, pages 167–172, Goteborg, May 2000.

111. T. A. Meynard and H. Foch. Multi-level conversion: high voltage choppers and voltage-source inverters. In *Power Electronics Specialists Conference, 1992. PESC '92 Record., 23rd Annual IEEE*, pages 397–403, Toledo, June/July 1992.

112. P. T. Krein and R. M. Bass. Autonomous control technique for high-performance switches. *IEEE Transactions on Industrial Electronics*, 39(3):215–222, June 1992.

113. Wai Lau and S. R. Sanders. An integrated controller for a high frequency buck converter. In *Power Electronics Specialists Conference, 1997. PESC '97 Record., 28th Annual IEEE*, volume 1, pages 246–254, St. Louis, MO, June 1997.

114. B. Acker, C. R. Sullivan, and S. R. Sanders. Synchronous rectification with adaptive timing control. In *Power Electronics Specialists Conference, 1995. PESC '95 Record., 26th Annual IEEE*, volume 1, pages 88–95, Atlanta, GA, June 1995.

115. O. Trescases, Wai Tung Ng, and Shuo Chen. Precision gate drive timing in a zero-voltage-switching DC-DC converter. In *Power Semiconductor Devices and ICs, 2004. Proceedings. ISPSD '04. The 16th International Symposium on*, pages 55–58, May 2004.

116. S. Mapus. Predictive gate drive boosts synchronous DC/DC power converter efficiency. *Technical Report, Appl. Rep. SLUA281, Texas Instruments*, 2003.

117. J. Kimball and P. T. Krein. Continuous-time optimization of gate timing for synchronous rectification. In *Circuits and Systems, 1996., IEEE 39th Midwest symposium on*, volume 3, pages 1015–1018, Ames, IA, August 1996.

118. A. V. Peterchev and S. R. Sanders. Digital loss-minimizing multimode synchronous buck converter control. In *Power Electronics Specialists Conference, 2004. PESC 04. 2004 IEEE 35th Annual*, volume 5, pages 3694–3699, June 2004.

119. V. Yousefzadeh and D. Maksimovic. Sensorless optimization of dead times in DC-DC converters with synchronous rectifiers. In *Applied Power Electronics Conference and Exposition, 2005. APEC 2005. 20th Annual IEEE*, volume 2, pages 911–917, March 2005.

120. G. Schrom, P. Hazucha, J. Hahn, D. S. Gardner, B. A. Bloechel, G. Dermer, S. G. Narendra, T. Karnik, and V. De. A 480-MHz, multi-phase interleaved buck DC-DC converter with hysteretic control. In *Power Electronics Specialists Conference, 2004. PESC 04. 2004 IEEE 35th Annual*, volume 6, pages 4702–4707, June 2004.

121. R. Miftakhutdinov. Analysis and optimization of synchronous buck converter at high slew-rate load current transients. In *Power Electronics Specialists Conference, 2000. PESC 00. 2000 IEEE 31st Annual*, volume 2, pages 714–720, Galway, June 2000.

122. I. Babaa, T. Wilson, and Yuan Yu. Analytic solutions of limit cycles in a feedback-regulated converter system with hysteresis. *IEEE Transactions on Automatic Control*, 13(5):524–531, October 1968.

Index